MW00528582

"The subtitle of Jeremy Rifkin's new breakthrough book 'Rethinking our Home in the Universe' is an appropriate description of this amazing masterwork. Rifkin is asking us to literally rethink everything we know and do. His premise is that we have misunderstood the planet we live on over all of history. We've long believed that we live on a land planet when in reality we live on a water planet. And that rethink requires an entirely new playbook for how we live and thrive in the future. The book delivers a multidiscipline tour de force for rethinking every aspect of our lives: our concepts of time and space; the way we organize economic life; how we educate our children; and how we relate to nature. Rifkin has dared to suggest the ultimate heresy – asking us to rebrand our orb in the cosmic theater. We live on Planet Aqua."

–Jerry Wind, Professor Emeritus, The Wharton School, and Founding Director of the Joseph A. Lauder Institute at The Wharton School

"When humans decided to tame the vast waters of our planet for the exclusive use of our species, it set in motion a time-bomb and now a somber day of reckoning. Rifkin's *Planet Aqua* asks us to listen to the cry coming from the very heart of nature warning us that the hydrosphere, not the human-sphere, is the prime mover of life on our small corner of the universe. Rifkin has given us a powerful blueprint for rethinking our home, reminding us we live on a water planet not a land planet, and that changes everything if only we heed him."

–Maude Barlow, Chairperson of the Board of Directors – Food & Water Watch

"Rifkin offers a much-needed wake-up call to rethink the very nature of our existence on a water planet. The stakes of forgetting this primordial human relationship with water are high. This book surveys the full scope of human interactions with water, diving into the ideas that have given rise to urban hydraulic civilization around the world and the new ways we will need to adapt in the throes of a rewilding planet. Rifkin offers a hopeful reorientation of our relationship to the hydrosphere that can help redirect humanity toward a safer and more prosperous future."

–Ani Dasgupta, President & CEO, World Resources Institute

"Rifkin's *Planet Aqua* is an enlightening read. Like a timely beacon in the darkness, the book offers an urgently needed guide for our lost species confronted by a rewilding hydrosphere. In oriental Asia, 'water is life' is the centerpiece of our cultures and shapes our behavior and relationships, defining our very existence. In Asian philosophy, poetry, and mythology, water is where the ancient sages find peace and freedom. Seeing with water and thinking with water is 'the highest virtue.' It is time for a great turnaround by relearning how to adapt our species to the hydrosphere rather than adapting the hydrosphere to the utilitarian whims of our species. Rifkin's *Planet Aqua* guides us on that journey."

–Changhua Wu, Chair, Governing Council of the Asia Pacific Water Forum

"In *Planet Aqua*, Rifkin reminds us that the waters animate all of life on the Blue Planet. As he writes 'To "be of the waters" is to steady our course and secure our place in a complex living organism, of which we are but one of the players, whose own resilience and well-being depends on "going with the flow."'"

–Vandana Shiva, Environmental Activist and Recipient of the Right Livelihood Award

"Jeremy Rifkin's *Planet Aqua* is a paradigm shift when we suddenly realize that we live on a water planet and not a land planet. Rifkin digs deep into our species' history, asking what is truly valuable to our existence and our fellow creatures and concludes with a simple observation that it is all about how we distribute real wealth: water for all life. I loved *Planet Aqua*."

–Gordon Gill, Principal, Adrian Smith + Gordon Gill Architecture (ranked the number one architecture firm in the United States by Architect *magazine)*

Planet Aqua

To Carol L. Grunewald
For a lifetime of sharing ideas.

Jeremy Rifkin

Planet Aqua

Rethinking Our Home in the Universe

polity

First published by Polity Press in 2024

Polity Press
65 Bridge Street
Cambridge CB2 1UR, UK

Polity Press
111 River Street
Hoboken, NJ 07030, USA

ISBN-13: 978-1-5095-6373-9 (hardback)

A catalogue record for this book is available from the British Library.

Library of Congress Control Number: 2023951427

Typeset in 10.5 on 14pt Sabon LT Pro
by Cheshire Typesetting Ltd, Cuddington, Cheshire
Printed and bound in Great Britain by CPI Group (UK) Ltd, Croydon

For further information on Polity, visit our website:
politybooks.com

Contents

Detailed Contents

Part IV Sublime Waters and a New Ontology of Life on Earth

Acknowledgments

I'd like to thank Claudia Salvador and Daniel Christensen for their keen editing of *Planet Aqua*. Their intellectual insights, command of the language, and dedication to the mission, were invaluable. I'd also like to thank our research director Michael Ricciardi for aligning the vast data that we worked with to make sure that our mission stayed on track . . . a great and dedicated team.

A special thanks to John Thompson and Elise Heslinga at Polity Press for jumping in at a moment's notice to speed the final manuscript to press. Their enthusiasm, acute editorial suggestions, and deep commitment to the project have kept us buoyed.

Introduction

What would happen if we were to awaken one day – all eight billion of us – only to realize that the world we live in and experience, and to which we are deeply attached, suddenly appeared eerily alien, as if we'd been teleported to some other distant planet where identifiable markers by which we've come to understand our existence were simply missing – no less our sense of agency? That frightening prospect is now. Everything we thought we knew about our home in the universe now seems so misbegotten. All the familiar signposts to which we've been attached, and which gave us a sense of belonging and direction, seem to have disappeared in a puff. In their wake, it's we who feel dispossessed and lost on our own planet. And each of us, in our own way, is frightened and unable to conjure up options as to where to turn for solace and engagement.

What's occurred to make us feel like aliens in our own small orb of the universe? It's difficult to hear this, but for a long time – at least the past six millennia of what we've come to identify as "civilization" – we misjudged the very nature of our existence and to what we owe our lifeline. To put it bluntly, our species, particularly in the Western world, has come to believe that we live on *terra firma*, a verdant green expanse of solid ground upon which we stand and thrive and which we call our home in the cosmic theater. That sense of place was shattered on December 7, 1972.

The crew of the Apollo 17 spacecraft on its journey to the moon took a quick photo of the Earth from a distance of 29,400 kilometers, showing in vivid detail a beautiful blue orb illuminated by the sun, changing the very way humanity would come to perceive our home. The long-held vision of a verdant and green *Earth* was instantly reduced to a veneer atop of which has always been a water planet circling the sun and, to date, seemingly alone in its multiple shades of blue in our solar system and perhaps in the universe. On August 24, 2021, the European Space Agency introduced the term "Planet Aqua." America's National Aeronautics and Space Administration (NASA) concurs, saying on their website that "looking at our Earth from space, it is obvious that we live on a water planet."

Of late, our water planet has become the center of attention around the kitchen table, in our neighborhoods and communities, in the halls of government, and in industry and civil society. The reason is that the planetary hydrosphere is rewilding in ways that were seemingly unthinkable just a few years ago, taking us into the early stages of the sixth extinction of life on Earth. Our scientists tell us that upwards of 50% of the species on Earth are now threatened with extinction within the next eighty years, many in the lifetime of today's toddlers.[1] These are species that have inhabited the Earth for millions of years.

Our climate is heating up from the emission of global-warming gases into the atmosphere – CO_2, methane, and nitrous oxide. Here's why the planet's hydrosphere is affected. Every one degree Celsius rise in temperature on Earth from global-warming emissions results in faster evaporation of water from the ground and the sea into the atmosphere and a seven percent increase in the concentration of precipitation in the clouds and more violent and exponential water events – frigid winter weather that brings with it powerful atmospheric rivers and blockbuster snows, massive spring floods, prolonged summer droughts, deadly heatwaves and wildfires, and catastrophic autumn hurricanes and typhoons – each devastating the planet's ecosystems with the loss of human life and the lives of our fellow creatures, along with destruction of the societal infrastructure.[2]

Take a look at the following short list of the damage inflicted thus far. It's sobering to read but needs to be vetted if we are to shake ourselves from either denial, somnolence or, worse still, despondency.

- Today, 2.6 billion people experience high or extreme water stress. By 2040, a total of 5.4 billion people – more than half of the world's projected population – will live in 59 countries experiencing high or extreme water stress.[3]

- 3.5 billion people could suffer from water-related food insecurity by 2050, which is an increase of 1.5 billion people from today.[4]
- Over the past decade, the number of recorded water-related conflicts and violent incidents increased by 270% worldwide.[5]
- One billion people live in countries that are unlikely to have the ability to mitigate and adapt to new ecological threats, creating conditions for mass displacement of populations and forced climate migrations by 2050.[6]
- Flooding has been the most common natural disaster since 1990, representing 42% of recorded natural disasters. China's largest event was the 2010 floods and landslides, which led to 15.2 million displaced people. Flooding has also dramatically increased in intensity across Europe, accounting for 35% of recorded disasters in the region, and is expected to rise.[7]
- Droughts, heatwaves, and massive wildfires are spreading across every continent, ravaging ecosystems and destroying infrastructure around the world.
- In the late spring of 2022, "severe to extreme drought" affected 32% of the contiguous United States.[8] As of 2023, 1.84 billion people – nearly 25% of humanity – were living in countries experiencing serious drought. 85% of people impacted by drought reside in low or middle-income nations.[9]
- Record temperatures ranging from 110 to 122 degrees Fahrenheit are being reached in regions across the planet. Death Valley in California recorded a temperature of 130 degrees Fahrenheit on July 9, 2021.[10] Even Antarctica set a world record of 65 degrees Fahrenheit in an extraordinary heatwave in April 2021. The years between 2015 and 2021 were the warmest on record.[11] This record was shattered two years later in July 2023 when the planet experienced the three hottest days in a row ever documented.[12]
- Within the first nine months of 2023, 44,011 wildfires had burned 2,342,143 acres of land in the United States.[13] The U.S. forest fires were eclipsed in 2023 with the burning down of 45.7 million acres of boreal forests in Canada in just a six-week period.[14] These forests contain 12% of the world's land-based carbon, which, if burned to the ground, is equivalent to 36 years of global carbon emissions from burning fossil fuels.[15]

- Smoke pouring down from Canadian wildfires created air quality so bad that the sky over New York turned bright orange and the air quality was deemed the worst on Earth, followed by Chicago, Washington DC, and other cities, and millions of people were told to remain indoors.
- Nineteen countries are at risk of rising sea levels, where at least ten percent of each country's population could be affected. This will have significant consequences for low-lying coastal areas in China, Bangladesh, India, Vietnam, Indonesia, and Thailand over the next three decades – as well as cities with large populations like Alexandria in Egypt, The Hague in the Netherlands, and Osaka in Japan.[16]
- By 2050, 4.7 billion people will reside in countries with high and extreme ecological threats.[17]
- Scientists reveal the alarming discovery that the melting of polar ice sheets and mountain glaciers and the extraction of an unprecedented volume of water pumped out of the ground for irrigation and human consumption have changed the way mass is distributed across the planet and altered the Earth's spin on its axis with untold implications for the future of life.[18]
- Oxygen levels in the oceans have plummeted in the wake of climate change by as much as 40% in some regions.[19]
- By 2050, 61% of all the hydropower dams on the planet will be in river basins at "very high or extreme risk of droughts, floods, or both."[20]
- Twenty percent of all the remaining fresh water on Earth is in the five interlinking Great Lakes of North America.[21]
- The World Bank reports that "in the last 50 years, fresh water per capita has fallen by half."[22]

If we are the despoilers of life, we might at long last also be the saviors – maybe. There are reasons to be guardedly hopeful, although not naively expectant. But this dramatic turnaround will depend on a change in how we conceive of human agency and our relationship to the planet. We will need to undertake a powerful post-mortem on how it came to be that our species broke rank some six millennia ago from all of the other living creatures inhabiting the Earth, who, from time immemorial, adapted moment-to-moment to an animated and ever-evolving planet.

Our earliest ancestors were animists and conceived of the world around them as alive, vibrant, and brimming with spirits continually interacting in a boundaryless nature, of which our species' agency was intimately intertwined. The big turnaround came when our forebears began to use their extraordinary mental acumen and physical dexterity to change course and adapt all of nature to the utilitarian caprice of our species, at the expense of the depletion of the natural world.

Six millennia ago, along the Euphrates and Tigris Rivers[23] in what is now Turkey and Iraq and, shortly thereafter, the Nile River in Egypt,[24] the Ghaggar-Hakra and Indus rivers in the Indus Valley, which today covers parts of India and Pakistan,[25] and the Yellow River in the Huang He Valley in China[26] and later across the Roman Empire,[27] our forebears began to harness the planetary waters for the exclusive use of our human family. They constructed elaborate dams and artificial reservoirs, erected dikes and levees, and carved out canals across the great rivers, sequestering, propertizing, and commodifying the waters for the use of their populations, fundamentally altering the natural ecology of their bioregions. These hydraulic infrastructures gave rise to what historians call urban/hydraulic civilization. The capturing of waters in regions around the world continued unabated, reaching a high-water mark in the early decades of the twenty-first century.

Although scant attention has been given over to this extraordinary reorientation of the planetary hydrosphere by historians and anthropologists, and even less economists and sociologists, the fact is that the emergence of dense urban life is an inextricable derivative of the hydraulic infrastructure whose sole purpose is to serve the needs of only our species.

Much of the sociology, economy, and governance that has accompanied the human journey over these six millennia is enshrined in hydraulic infrastructure. While pockets of our species have remained outside these massive infrastructure bubbles – some until the present era – it's the great urban hydraulic civilizations that mark much of what we've come to regard as the historical footprint of human history.

Now, in the grips of a warming planet, brought on in large part by a fossil fuel-driven water/energy/food nexus, the urban hydraulic civilization is collapsing in real time. This vacuum is giving rise to a great reset in our species' relationship to the planetary hydrosphere. We are just beginning to relearn how to adapt our species to the requisites of an animated and ever-evolving and self-organizing planet, in which the hydrosphere plays the central role as the orchestrator of life. What we're talking about is a new form of neo-animism, driven by a sophisticated science-based and technologically deployed rapprochement with our aqua home.

Finding our way back on a fast-rewilding planet has triggered what our philosophers call the "sublime." The term was first introduced in 1757 by the Irish philosopher, Edmund Burke, who penned an essay entitled "A Philosophical Inquiry into the Origin of our Ideas of the Sublime and Beautiful," thereafter taken up by the philosophers of the day to become a centerpiece of the Enlightenment and Romantic Period, and later, in the nineteenth and twentieth centuries, the Age of Progress.

Burke describes the feeling of utter terror one feels in the presence of an overwhelming force of nature: it can be an imposing mountain range, a deeply sculptured canyon, a massive wildfire, a raging flood or hurricane, a line of deadly tornados touching down, an erupting geyser, a red-hot volcano spewing ash across the land, or a powerful earthquake prying open the very seams of the planet and swallowing everything in its wake. If viewed from a safe distance, and from harm's way, one's terror turns to a sense of "awe" in the presence of nature unleashed. Awe, in turn, leads to "wonder" about the powerful forces of nature and sparks our "imagination" about the meaning of existence, and, on occasion, leads to a "transcendent experience" – a reorientation of our sense of attachment to place in the bigger scheme of things on our shared planet.

This experience of the sublime has spawned a heated debate between two very different schools of philosophy about the meaning of life and our own personal relationship to existence. As our human family edges ever closer to an environmental abyss and the potential extinction of life on Earth, the question of how to respond to the sublime of nature has become a crossroad. It's here where each of us in our own way is beginning to ask the big question: which approach will take us to a transcendent experience, and will that experience be in the form of a new utilitarianism or a mindful and empathic biophilic reattachment to our home in the universe?

The Enlightenment philosopher Immanuel Kant argued that at the juncture where the trail of terror, followed by awe, wonder, and imagination in the throes of the sublime, meet, the "rational mind" – a non-material force, independent from nature's tempest and even immune to its overwhelming agency – steps in and takes over, using cold, detached, objective reason to subdue, capture, sequester, and tame nature's excesses, in order to meet the utilitarian needs of our human family. In short, human reason neuters the beast.

The philosopher Arthur Schopenhauer, however, would have none of Kant's rational detachment. He argued that although the sublime at first conjures up terror and a sense of helplessness in the eyes of the beholder, shortly followed by the triggering of one's sense of awe, wonder, and imagination when enveloped by powerful planetary forces, it can lead to a different path to transcendence – a feeling of compassionate belonging on an animated planet in which each individual is an agent and participant, enveloped by the irreducible unity of all of life that makes up existence.

Today, that standoff between cold objectivity and detached utilitarianism versus engaged biophilic attachment is playing out on every front as the forces of AI, technological singularity, and the metaverse square off against the forces of an enlightened neo-animist reattachment. The issue at hand is whether our species will continue to bend nature to our rational will or embrace nature's calling and rejoin the community of life on Planet Aqua. Where does all this come down and around? When we come to understand that we live on Planet Aqua. It's our milieu – the medium in which we dwell. The hydrosphere is not a thing but, rather, the animating force of the story of life on the planet. It's the driver of the other three principal spheres on Earth – the lithosphere, atmosphere, and biosphere – and the incubator of all of life yet to come.

Getting our ontology right – the nature of our being – is of the first order. Putting ontology into practice is of the second order. The chapters and pages that follow are in the way of a story and an account of where we've been, which has taken us and our fellow creatures to the brink of extinction, and where we will need to go in a "new order of the ages," *novus ordo seclorum*, that's just beginning to take wing. This new narrative and accompanying journey might give our species and fellow creatures a second lease of life on Planet Aqua.

None of the mentions in the chapters and pages that follow is theoretical but, rather, they are all phenomenologically grounded in the real-life experiences that have taken our species to this point in history on the planet. How a new "sublime" might unfold is an unknown that awaits the kind of future we embrace, subject to the butterfly effect and the good fortunes and misgivings that come our way with life on Planet Aqua.

What the data tell us is that we are on the cusp of an imminent collapse of urban hydraulic civilization after six millennia of history. The warming of the planet has unleashed a long-sequestered hydrosphere. Our water planet is evolving in wholly new ways that we can barely fathom. Ecosystems everywhere are collapsing, infrastructure is being devastated, and the lives of our species and our fellow creatures are increasingly at risk. The entire human-built environment is now a stranded asset and will have to be rethought, reimagined, and redeployed in new ways.

Consider, for example, a new study conducted by researchers at Northwestern University in the United States and published in 2023.[28] This study found that global warming is heating up and deforming the ground below the surface, meaning buildings, water and gas pipelines, electric utilities, subways, and other underground infrastructures are increasingly at risk. Chicago, for example, is already beginning to experience the early stages of sinking infrastructure, threatening the iconic building stock that made the city an architectural landmark in the twentieth century. Other urban megalopolises on every continent will inevitably sink, and not over a millennium of history, but more likely over the next 150 years or so, posing a potential winnowing of urban life, at least as we've come to know it.

As the saying goes, crisis creates opportunity. Our species finds itself mired in the greatest crisis since we first inhabited the planet – a mass-extinction event taking place in real-time. Rethinking every aspect of where we've come from, what we've believed in, how we've lived, and where we need to go to adapt and reaffiliate on Planet Aqua, whose hydrosphere is transforming itself in novel ways, is the business before us.

That process is already well underway. Urban "hydraulic civilization" is collapsing in various regions in real-time, while "ephemeral society" cued to an unpredictable hydrosphere is emerging in bits and pieces and just beginning to scale. If the former is characterized by long periods of sedentary life and shorter periods of migratory existence, the latter is wound around longer periods of migratory life and shorter periods of sedentary existence.

With this epoch shift comes a whole new vocabulary to accompany the processes, patterns, and practices that ride alongside a rewilding hydrosphere on Planet Aqua. Innovative affordances are flooding into the public square. The "slow water movement," "water internets and microgrids," "sponge cities," "water calendars," and the demarcation of the "water year" are among the numerous new terms becoming part and parcel of our daily lives, realigning our species' comings and goings on Planet Aqua. Communities are even introducing the new term "depaving," also known as "desealing" – a process of dismantling impermeable surfaces across neighborhoods and communities and replacing them with permeable green spaces, in order to free water and allow it to seep into the ground and, from there, follow its natural pathways. Solar and wind-generated "alchemic desalination" of saltwater is also trending. It is estimated that by 2050 more than a billion people on Earth will be drinking desalinated water produced by solar-powered osmosis. Already, portable suitcase-sized osmosis devices using less electrical power than cell phones are coming onto the market and becoming an indispensable accoutrement in a drought-ridden world.[29]

Now, an incipient neo-nomadism is emerging in the throes of a warming climate, transforming the very notion of infrastructure. Popup cities, tear-down and recycled infrastructure, poured-out 3D-printed temporary habitats, scaled vertical AI-driven indoor agriculture, including insect farming, using 250 times less water than conventional outdoor farming, are changing the socioeconomic landscape. Mapping of migratory corridors are now nascent, and the issuing of millions of climate passports are likely to follow and scale, as our species increasingly abandons conventional highly dense urban habitats caught up in global warming. A rewilding hydrosphere is fast changing the settlement patterns of our population on every continent, letting the waters determine how our human family is distributed, along with our fellow creatures.

National governments, long locked into sovereign states and fixed borders, are being challenged as well by "bioregional governance" across political boundaries, as local communities begin sharing the responsibility of stewarding their common ecosystem. Globalization, already dwindling as the climate warms and climate disasters undermine logistics chains and trade across ocean and air corridors, is giving over to glocalization as new more agile high-tech small and medium-sized enterprises (SMEs) and cooperatives in regions begin to engage directly with one another across digitally driven provider–user networks, at near zero marginal cost, bypassing traditional seller–buyer capitalist markets.

"Geopolitics," now senescent but fighting a torturous endgame to the ruin of the world, is being challenged – if still only lightly – by a fervent new "biosphere politics," as our human family comes to the realization that all of us and our fellow creatures live in an overarching biosphere that we have to share. The shift to biosphere politics, in turn, is being accompanied, if only tentatively, by the transition from military preparedness to defend political boundaries and property to climate-disaster rescue, recovery, and relief missions across shared ecosystems.

The rise of ephemeral society also comes with new terms to describe economic life. Ecological economics, which takes its cues from the First and Second Law of thermodynamics, edges us to a hybrid economic system only partially cued to what we call market capitalism, while increasingly entwined in an economy where finance capital becomes increasingly less important than ecological capital. Hydroism becomes the new rallying point in an ephemeral society.

On Planet Aqua, efficiency gives way to adaptivity as the primary temporal value, productivity becomes far less important than regenerativity, Gross Domestic Product (GDP) is pushed aside to make room for Quality-of-Life Indicators (QLI), and zero-sum games lose currency, while the network effect increasingly becomes the norm.

Learning to live on Planet Aqua comes with new ways to measure performance. Understanding the "water–energy–food nexus" and using the "virtual water index" to gauge how water is distributed both domestically and for export and import become the gold standards and as important as the carbon footprint in commerce and trade on a water planet.

Perhaps most encouraging, the collapse of urban hydraulic civilization and the rise of ephemeral society is being accompanied by a growing recognition of the "rights of waters" as the prime mover and animating force of all of life on Planet Aqua. Countries have begun enacting laws guaranteeing the legal rights of oceans, lakes, rivers, and floodplains to flow at will, as the hydrosphere rewilds in search of a new balance, and they are backing up these rights with the authority to bring infringements to court for adjudication.

These vast changes in the way we've come to think of our planet in the cosmic theater reboot the human story, taking us into a new life-affirming future. We live on a water planet and every aspect of our existence flows from this incontrovertible truth. Rebranding our home in the universe as Planet Aqua, and introducing this second naming of our planet into government constitutions, bylaws, codes, regulations, and standards, is the first giant step toward a realignment with the waters that animate our very existence. This epiphanic moment marks the beginning of a new transcendent journey to rekindle the pulse of life on our water abode. These and countless other choices we make over the course of the next several generations will determine whether life on the planet will be re-endowed and our species renewed. There is only a single agenda before us: making peace with a rewilding hydrosphere and finding new ways to flourish along with our fellow creatures. All else is a distraction.

PART I

The Imminent Collapse of Hydraulic Civilization

CHAPTER 1

First There Was the Waters

The great mystery in the history of Earth is how did life come to be? The first tantalizing clue appears early in the opening lines of the Book of Genesis in the Bible. Shlomo Yitzchaki, also known as Rashi, was a renowned French rabbi of the eleventh century, whose commentary on the Talmud remains an authoritative interpretation of biblical script. The biblical account of the Creation, notes Rashi, begins with the startling admission that first there were the waters, which preceded God's creation of heaven and earth.[1] Genesis leads off with a passage that suggests that in the beginning the Earth was formless and empty, and darkness was over the surface of the deep. And the spirit of God was hovering over the waters.[2]

God then parted the primordial waters, creating heaven and earth and day and night, separating land from the oceans, and populating the Earth with every living creature. His last and most prized creations were Adam and Eve, made in God's image, and fashioned from the dust of the Earth. To be fair to historical accounts, the Genesis story of the waters existing before the Creation was not a lone rider. The earlier Babylonian civilization tells a similar story of creation. Other creation stories from around the world follow along the same lines. The ancient stories of primordial waters are recently gaining interest, as scientists begin to unwrap the secrets of the formation and evolution of the universe and our own solar system and planet and to look at the role water plays in the unfolding of the cosmos.

These narratives of the Earth's beginnings, which place the waters before the Creation, have taken on existential importance because of the tumultuous changes taking place in the Earth's hydrosphere. While global warming, resulting from the industrial emissions of CO_2, methane, and nitrous oxide from the burning of fossil fuels, affects all four of the Earth's primary spheres – the hydrosphere, lithosphere, atmosphere, and biosphere – its biggest impact is on the Earth's hydrosphere. The Earth's ecosystems, which have developed over a mild-weather regime during the Holocene era of the past 11,000 years, are collapsing in real time with the onslaught of climate change and the rewilding of the waters, taking the planet into the sixth extinction of life on Earth. (The last time the Earth experienced a mass extinction of life was 65 million years ago.)

It's no wonder the scientific community is in a frantic search to understand the intimate workings of the Earth's hydrosphere and its impact on the lithosphere, atmosphere, and biosphere, in order to better adapt to the changing ocean currents and the Gulf Stream; the effects of the melting of the last remnants of the previous Ice Age on land and sea; the disruption and shifting of the Earth's tectonic plates; the unleashing of earthquakes emanating from the Earth's mantle; and the sharp increase in potential eruptions from thousands of previously dormant volcanoes.

If there were any doubt as to the overwhelming agency of the waters in directing the planet, consider the scientific findings that show that the way the waters are distributed alters the very axis of the Earth – its tilt.[3] And that's exactly what has been happening since the 1990s on our planet. The reason is that the warming of the planet from climate change is quickly melting the last remaining glaciers and ice sheets of the Pleistocene in the Arctic region. The massive volume of water released is spreading across the oceans and altering the way the weight of the planet is distributed, and changing the planet's spin on its axis.[4]

New research has also found that the recent pumping of groundwater for agriculture to feed a growing population, which now tops eight billion people, is also contributing to the way the waters are distributed – "enough to make the planet's axis shift." In India in 2010 the population pumped 92 trillion gallons of water from underneath the ground. While the change in the tilt of the Earth occasioned by human-induced climate change is only likely to slightly "alter the length of the day by a millisecond or so over time," it's enough to be awed by the agency of the waters and their impact on the planet.[5]

The question that scientists are asking is where did the primordial waters come from and how are they constituted? Astronomers have long entertained the notion that water is ever present across the universe, and journeyed to a newly formed Earth 3.9 billion years ago by the heavy bombardment of comets composed mainly of ice. New studies, however, favor a second complementary source of the waters – leaching from molten rock deep under the surface of the Earth.[6] Recent findings also suggest that the ancient Earth may have been a water world without continents, reinforcing the biblical description of the primordial waters that preceded God's creation of land.[7]

Although the scientific community has yet to fully decipher the connection between the waters and the evolution of life on Earth, it's a fact that every species is comprised primarily of water from the hydrosphere. All of this takes us back to Adam in the Garden of Eden, who was long thought to have been fashioned from the dust of the Earth. Actually, water makes up much of the composition of sperm, and the human body is gestated in water in the womb. In some organisms, upward of 90% of their body weight comes from water and, in humans, water makes up approximately 60% of an adult body.[8] The heart is about 73% water, the lungs are 83% water, the skin is 64% water, the muscles and kidneys are each 79% water, and the bones are 31% water.[9] Plasma, the pale-yellow concoction that transports blood cells, enzymes, nutrients, and hormones, is 90% water.[10]

Water plays an essential role in managing the intimate aspects of living systems. The list of particulars is impressive. Water is:

> A vital nutrient in the life of every cell [and] acts first as a building material. It regulates our internal body temperature by sweating and respiration. The carbohydrates and proteins that our bodies use as food are metabolized and transported by water in the bloodstream. [Water] assists in flushing waste mainly through urination. [Water] acts as a shock absorber for [the] brain, spinal cord, and fetus. [It] forms saliva [and] lubricates joints.[11]

Water is flowing in and out of our bodies every twenty-four hours. In this sense, our semi-permeable open systems bring fresh water from the Earth's hydrosphere into our very being to perform life functions, after which it is returned to the hydrosphere. If there's a case to be made that the human body – and all other living creatures – is more like a pattern of fluid activity than a fixed structure and operates as a dissipative system feeding off energy and excreting entropic waste, rather than a closed mechanism importing energy to secure its own autonomy, the cycling and recycling of H_2O is an appropriate point of departure.

Every human being knows intuitively that water is life. We can go without food for up to three weeks, but we can only go without water for a week or so on average before the risk of perishing. But now, the hydrological cycle is convulsing in ways that we struggle to grasp, changing the dynamic of all of the other Earth spheres and the prospect of the survival of our species and fellow creatures. We've been there before.

Déjà Vu and the Second Deluge

The very beginning of human recollection in history, as told by peoples everywhere, is the story of the great flood that engulfed the Earth. While Western civilization tells of the great flood rained down by Yahweh that swallowed all of life, with the exception of Noah and his family and a male and female from each species taken aboard his ark, others recount their own tales of a great flood and survival of the Creation. In recent years, the scientific community has uncovered evidence of catastrophic floods in various regions of the world with the melting of the last Ice Age. In Eurasia, North America, and elsewhere, giant ice-dammed lakes melted, unleashing massive glacial floods, and once-frozen rivers spilled over their banks onto adjacent lands, taking creatures to their deaths. The devastation caused by the melting ice sheets burrowed into the collective psyche of our ancient ancestors and was the first shared historic memory passed on orally and later with the advent of script to present times.

Now, eleven millennia later, the hydrosphere is rebelling once again in the throes of global warming of the planet. Scientists are warning us that half of the species on Earth are threatened with extinction within the next eighty years.[12] These are species, many of whom have inhabited the Earth for millions of years. Our scientists are engaged in a heated debate as to what has precipitated the current extinction event. Most place the blame on the fossil fuel-based Industrial Age and the spewing of massive amounts of carbon dioxide, methane, and nitrous oxide into the atmosphere, leading to a warming of the climate – with ample evidence in the geological record to back up the claims. Others argue that the journey to extinction began as far back as the formation of the first great hydraulic civilizations in the Mediterranean, North Africa, India, China, and elsewhere starting around 4000 BCE.

For 95% of the history of *Homo sapiens*, our forebears lived as our fellow creatures, foraging and hunting, continually adapting to the changing seasons and nature's ebbs and flows.[13] The hominid species, culminating with *Homo sapiens* some 200,000 to 300,000 years ago, lived on a perilous planet punctuated by 100,000 years or so of glaciation and followed by 10,000 years or more of warming interludes. The last deglaciation of the Pleistocene some 11,000 years ago gave rise to the temperate climate we've known ever since. With the coming of the Holocene, our ancestors settled into a sedentary existence categorized by agriculture and the pastoralization of animals – the Neolithic Age. That period of history eventually morphed into the rise of urban/hydraulic civilization 6,000 years ago in the Mediterranean and, shortly thereafter, India and China. For the first time in the history of our species on Earth, we reversed course from always adapting to nature's flows like all the other creatures, to an abrupt turn around, adapting nature to our species' desires and designs. The rise of urban/hydraulic civilization over six millennia of history culminated in the fossil fuel-based Industrial Age, the rise of capitalism, the warming of the planet's climate, and a hydrosphere wreaking havoc on the Earth.

In the past decade the United States experienced 22 extreme climate-related weather events, all triggered by the radical reorientation of the hydrospheric cycle, each with more than a billion dollars in losses to the environment, economy, and society.[14] In 2021 alone, climate disasters with damages exceeding $145 billion swept the country, including a frigid cold wave in the South and Texas, massive wildfires that spread across Arizona, California, Colorado, Idaho, Montana, Oregon, and Washington State, a summer/fall drought and heatwave that engulfed the western U.S., monumental flood events in California and Louisiana, a series of scattered tornados, four tropical hurricanes, and seven other severe weather events. The total cost to society, the economy, and the environment by climate-change-related severe weather events between 2017 and 2021 – all related to a rapidly reorienting hydrological cycle – topped $742 billion, according to the National Oceanic and Atmospheric Administration.[15] To set the losses in perspective, the entire U.S. Bipartisan Infrastructure Plan, signed into law in 2021 to reduce global-warming emissions and build out a smart resilient Third Industrial Revolution infrastructure for the country over the coming decade, totaled only $550 billion for climate-related programs in the legislation. To bring the crisis closer to home, four in ten Americans live in a county ravaged by "climate-related disasters in 2021."[16]

Worse still, in the U.S., 43% of the population live in communities that rely on dams, abutments, reservoirs, and artificial reefs that are aged and in disrepair – the average dam in the U.S. is nearly sixty years old, and the average levee over fifty years old.[17] These ancient hydraulic infrastructures were not engineered to withstand the floods, droughts, heatwaves, wildfires, and hurricanes of an increasingly unpredictable hydrological cycle. America is not alone. A 2022 study published in the journal *Water* reported that by 2050, 61% of all the hydro-powered dams on the planet would be in river basins at "very high or extreme risk of droughts, floods, or both."[18] Every country faces a similar dilemma – continuous repair and build-back of hydro infrastructure, which is a losing game, or free the waters to run their course and establish a new equilibrium. If taking the latter approach, the U.S. and governments everywhere will need to create massive buyout programs and relocation initiatives to allow for the resettlement in areas out of harm's way, and let the waters rewild and spur new ecological thresholds for the flourishing of life.

We can't say we weren't warned. The great scientific minds who have raised the alarm over the past century include the renowned Russian geochemist Vladimir Vernadsky who, in the early twentieth century, described the biosphere and argued that the hydrosphere is the critical agency that dictates the evolution of life on Earth. Around that same time, Harvard biologist, physiologist, chemist, and philosopher Lawrence Joseph Henderson suggested that water may be the missing link of animating life on Earth and the cosmos.[19] More recently, the biologist Lynn Margulis, who, along with the chemist Sir James Lovelock, introduced the widely accepted hypothesis of the Earth as a self-organizing system – which they called Gaia – worried about the impact that human civilization is having on the planet's hydrosphere.[20] Vernadsky, Henderson, and Margulis were of the same mind, having come to the realization that the "waters" are the animating force of life on Earth and likely elsewhere in the universe. More recently, others from the fields of chemistry, physics, and biology have begun to unlock the previously unexplored qualities of water.

Our Aquatic Self: How Humans Emerged from the Deep

Planet Aqua tells a new story of how life emerged on our planet, with water as the "prime mover," altering the very way our species perceives itself and its relationships on the blue marble. Although our human family has become partially at ease with the idea that we descended from our close relatives, the primates, more unsettling is the recent scientific finding that our roots lie far earlier in the history of evolution, in the great depths of the oceans. It's a long-accepted dogma among paleontologists that our species' evolutionary origins wind back to the first micro-organisms that inhabited the great oceans, but there was little evidence to bolster the theory in the way of tracing our lineage.

Over the past several decades, however, biologists have begun to fill in at least some of the missing links that trace our species' evolution far back in the aquatic record. In 2006, Neil Shubin, a University of Chicago professor, reported in a pair of articles in the journal *Nature* that while chipping away at an ancient rock formation in the Bird Fjord on Canada's Ellesmere Island, his team had discovered the fossil remains of a creature that grew to nine feet or more in length, and lived 375 million years ago, at the moment in geological history when fish were evolving into the first four-legged animals known as tetrapods. These strange animals had fish scales, fangs, and gills, but also anatomical features found only in animals that spend some time on land. Shubin and his colleagues named this creature "fishapod."[21] The fishapod has a broad skull, flexible neck, and, like the later crocodile, eyes high up on the head, allowing it to see across water and toward the horizon. The creature also sported a large interlocking ribcage, which suggested lungs and breathing. Researchers speculate that the creature's trunk was strong enough to support it in the shallows or on land. The unexpected finding was when they dissected the creature's pectoral fins they discovered the beginnings of a tetrapod hand and a primitive version of a wrist and five fingerlike bones, leading Shubin to exclaim that "this is our branch. You're looking at your great-great-great-great cousin!"[22]

Then, in 2021, scientists broke 160 years of thinking about the evolution of life on Earth with findings published in the journal *Cell*.[23] Using genome mapping of primitive fish conducted at the University of Copenhagen and other research laboratories, they dispelled the conventional wisdom that around 370 million years ago, primitive lizard-like animals – the tetrapods – began to migrate onto land, with their fins morphing to limbs and organs that allowed for air breathing. The new genomic findings suggest that fifty million years before tetrapods came onto land, the fish carried the genetic codes for limb-like forms and primitive functioning lungs for air breathing. Humans and a type of ancient primitive fish that still swim in lakes and rivers, the bichir, even share an essential function in the cardio-respiratory system, the conus arteriosus, a structure in the right ventricle of every heart, which allows the organ to deliver blood to the body. These genetic codes in the human genome and a group of primitive fish are an extraordinary finding, showing that our species shares a genetic history with the fish that swam the oceans for hundreds of millions of years, connecting us to the ancient waters from which life has evolved over eons of history.[24]

Our species is "of the waters," from the very first moment of life in the womb. While 60% of the adult human body is of water, the fetal body consists of 70 to 90% water but diminishes toward the time of birth.[25] Rock paintings from 10,000 years ago show our ancient ancestors swimming in various poses familiar to us today – the breaststroke and doggy paddle. An Egyptian clay seal dated between 9000 and 4000 BCE shows Sumerians doing the crawl. References to swimming are found in the Epic of Gilgamesh, the very first literary work of record written between 2100 and 1200 BCE during the Mesopotamian civilization.[26]

Our history is one of continuous immersion in the waters, for both the purposes of survival and recreation. We drink, swim, dive, float, luxuriate, bathe, engage in baptismal rituals and renewal, commune with the spirit world of the deep, and harness the waters to govern economic and social life – in other words, we live in a water milieu, both inside and outside, from conception to death. And all the water molecules that comprise our liquid being – in the cells, tissue, and organs – journey on to other pools and realities during our lifetime and thereafter, taking up new residence. How strange then to hear the biblical account of life interpreted as a cycle "from dust to dust," when it's more accurate to describe daily life on the blue marble as an ever-shifting and repositioning from liquid milieu to liquid milieu.

Perhaps that genetic link with our ancestors of the deep runs even deeper but has gone largely unnoticed during our conscious waking moments. Many of our dreams are about water or have water as a central theme or metaphor guiding the imagination. Dreams with the water motif often address the intimate gamut of human emotions – the fears, hopes, tribulations, and expectations that lie tucked away in the unconscious and stir the imagination. Psychiatrists, for example, often interpret patients' dreams of drowning with a feeling of being overwrought, while dreams of being dunked into water conjure up a spiritual cleansing, rebirth, or renewal of life. The water metaphor in dreams allows the unconscious mind to intimately plumb the depths of the psyche – more so than any other medium – suggesting the entwined relationship that human beings have with the waters is possibly a memory embedded in the very sinew of our collective unconscious.

The historian Mircea Eliade captured our species' intuitive sense of the importance of water as the animating force of life. He wrote:

> Water symbolizes the whole of potentiality; it is *fons et origo*, the source of all possible existence . . . water is always germinative . . . in myth, ritual, and iconography, waters fill the same function in whatever type of cultural pattern they find it: it precedes all forms and upholds all creation . . . Every contact with water implies regeneration.[27]

It's often the case that the most important ways we come to see ourselves and communicate with others is reflected in how we use language. Although we are often unaware of how we rely heavily on metaphors to describe and convey our thoughts, they are the way we express ourselves. The iconic Roman Catholic priest, theologian, and philosopher Ivan Illich reminds us that "water has nearly unlimited ability to carry metaphors."[28] And it's here that we come to understand the prime importance we attach to the waters. No surprise that the waters that animate all of life come to play a key role in how we communicate our thoughts to others.

The water metaphors are ubiquitous across every culture and language. The list is endless: a flood of tears, break the ice, tip of the iceberg, a ripple effect, water under the bridge, feeling swamped, an icy stare, the idea is all wet, hook line and sinker, get your feet wet, mind in a fog, a frosty look, drowning in thought, troubled waters, sink or swim, on Cloud 9, in hot water, wet behind the ears, treading water, a watered-down proposal, fish out of water, a trickle of news, surfing the internet, riding the waves, etc.

Given that our species is land-based, it's not hard to understand that we've come to think of green space as our primal dwelling. Over millennia of history, the identification with landscapes has been our obsession, at least consciously. We've largely taken the water for granted, thinking of it as a resource instead of a lifeforce and a utility rather than the milieu within which we dwell.

In recent years, however, the various scientific disciplines have begun to turn their attention to "waterscapes," asking the question of the role they play in defining the nature of humanity. What they're beginning to find is that at the subconscious level we remain of the waters. Now, that reality is beginning to surface, in the throes of a warming climate, and a rewilding of the planet's hydrosphere. The atmospheric rivers, spring floods, summer droughts, heatwaves and wildfires, and fall hurricanes and typhoons have awakened us to the fact that like every other species we dwell on a water planet.

Our awakening is bittersweet. If there is an upside, it's that our biologists, ecologists, engineers, architects, urban planners, and others, are rediscovering our subliminal ties to waterscapes. Although still regarded as an appendage to landscapes, we're learning that our deeper affiliation still lies with the waters. Researchers are just now starting to ask how we human beings regard blue space relative to green space, and they are finding that our affinity to the former has always been there, even if masked over the millennia. First, consider where much of our population lives. Ten percent of our species live on coastlines and another 40% live within sixty miles of the coast.[29] Moreover, 50% of the world's population live within three kilometers of a freshwater body and less than ten percent of our species live further than 10 kilometers or 6 miles away.[30]

Now that the hydrosphere is becoming a wildcard in a warming climate, the therapeutic and health impacts of blue space are getting a second hearing among biologists, ecologists, urban planners, architects, and the public at large, even if the new studies approach the subject more as an extension of green space, when, in truth, it is the hydrosphere that generates the lithosphere and not the other way around.

A review of multiple studies on the impact of blue space on human health and well-being, conducted by researchers at the Institute for Hygiene and Public Health at the University of Bonn, discloses the aesthetic and health-related effects of being embedded in or on the waters. The phenomenological effects – i.e. the embodied experience – of engaging blue space were rich in content. When approaching blue space, respondents in the studies became acutely aware of the increase in humidity and the rich diversity of wildlife congregating along the periphery and over, on, and in the blue commons. The awakening of the senses had the effect of enveloping the subjects within a blue milieu. The colors, sound, clarity, motion, and context of the waters drew the subjects into an alternative realm far denser and more animated in comparison to the cacophony of banging and metallic smells and accompanying gassy emissions that come with a mechanistic urban sprawl.

Subjects in the various experiments reported a heightened sense of immersion in blue space and a feeling of acute aliveness amid the rich diversity of life spawned by the waters. In particular, the sounds, colors, and flows of the waters drew a range of exhilarating physiological responses and sparked deep feelings of oneness with the bluescapes.

The authors of the mega study describe multiple experiments in bluescape attachment and list some of the shared experiences of the subjects that took part: "People admire the sounds of water and great importance is attached to the variety and special nature of these sounds, ranging from calm, laminar flows to energetic, roaring sounds [and] consider the calm sounds of water to be restorative." The color of water also triggers different emotional reactions. Blue water, for example, is thought of as pure, while yellow water, often not so. Blue water is also associated with coolness, whereas white water with roaring sounds and power.[31]

Likely, everyone who has ever walked along an ocean coastline with the waves rushing onto shore, and just as quickly receding into the deep, can't help but be mesmerized with what might lie below and beyond. And who hasn't been moved by looking across the vast expanse of ocean that defies enclosure, letting each of us ponder the immensity of existence and ruminate about how this seemingly infinite space came to be, and how each of us fits into this bigger picture here on Earth.

Among the most interesting findings about our species' relationship to the waters is how every culture across history has sought aquatic environments for restorative purposes. A study done on patients' and students' preference of pictures to hang on the walls found that "aquatic scenes consistently ranked highest."[32] Although water is both the medium and milieu in which all of life exists on our planet, we are often oblivious to the critical role it plays in animating the Earth's other great spheres and buoying our own prospects as a species. The late American novelist David Foster Wallace shared a parable that goes to the core of our intimate but little-examined relationship with the waters. It goes like this:

> There are these two young fish swimming along and they happen to meet an older fish swimming the other way who nods at them and says "morning boys. How's the water?" And the two young fish swim on a bit and then eventually one of them looks over at the other and says, "what the hell is water?"[33]

Water is omnipresent and guides every moment of our physical existence as well as our embodied relationships in the world, although often undetected since it's the medium in which we dwell. In a study prepared by Shmuel Burmil, Terry Daniel, and John Hetherington in the *Journal of Landscape and Urban Planning,* the authors share some thoughts on water's ever presence. The study begins with water's reflective nature. Water, they write:

> is capable of almost complete return of light waves from its surface . . . when the surface is calm, extremely clear images of mountains, rocks, trees, wildlife, and at times the observer him/herself are displayed . . . the color of water is also influenced by the eroded material suspended in it. The Colorado ("red") River is named after the color of mud it carries . . . Sound is created by water flowing down cascades over or around obstacles and by fish and other animals moving through the surface . . . There are the very subtle sounds of single drops falling and hitting the water surface, the rushing sounds of rapids or the thundering roar of a waterfall . . . In nature, water fills valleys, forms into pools, and meanders through arroyos . . . Water can be calm and reflective, lying still . . . or it can be in energetic motion forming rough vertical or angular planes

> . . . Water dramatically influences and shapes the landscape; it can create 'monumental, sculptured environments [as with the Grand Canyon in Arizona].'[34]

Water as a medium is ever-present, and if taken for granted, it's because, as Burmil et al. point out, "water has no form of its own [but rather] it takes the shape of its container."[35] In short, water animates all of life.

Geophysicists have long mused on whether there might be water down in the interior mantle and the core of the Earth. The mantle takes up 84% of the planet's volume, while the core accounts for an additional 15% of the volume of the Earth. The crust makes up the remaining one percent.[36] When it comes to the planet we live on, until very recently we knew precious little about what is going on down under. There was some speculation in the 1990s that the mineral ringwoodite – the principal component of what is known as the transition zone between the planet's upper and lower mantle, between 250 and 410 miles below the surface – might be storing massive amounts of waters for hundreds of millions or even billions of years.[37] Although at these depths and under extreme temperature, geophysicists think that the water had broken apart into its constituent elements of hydrogen and oxygen and came to be chemically embedded in the rock's crystal structure. Irrespective of the form it takes, it's still water.

Then, in 2014, Graham Pearson, a University of Alberta geologist unearthed a tiny diamond formed in the transition zone in Brazil and inside discovered a fleck of ringwoodite trapped there containing approximately one percent water – raising the question of how much water might be hidden under the Earth in its mantle and core.[38] At that time, Steven Jacobsen, a geologist teaching at Northwestern University and Brandon Schmandt, a University of New Mexico geologist, teamed up and began surveying the Earth's mantle using a dense network of 2,000 seismometers across the United States and found a massive amount of molten material in the transition zone below, making it crystal clear, according to Jacobsen, that the transition zone is full of water.[39]

Since then, geologists have discovered more waters both above and below the transition zone. In 2016, Jacobsen's team obtained a tiny diamond brought up from 600 miles below the Earth's surface in the lower mantle from a volcanic eruption ninety million years ago. The diamond contained ferropericlase, a key component of the lower mantle and that is less than one percent of water by weight.[40]

Given that the lower mantle makes up half of the Earth's mantle, Jacobsen and other geologists believe that another ocean mass is likely distributed between layer upon layer of rocks. Extrapolating from the data, Jacobsen suggests that in the entirety of the world's subduction zones that "we have one ocean mass in the oceans [and] another in the upper mantle . . . Let's suppose there are two more in the transition zone." Jacobsen speculates that there may be still another ocean mass in the crust and lower mantle combined, "making five ocean masses altogether."[41]

And these revelations about oceans beneath the oceans are far more than an esoteric curiosity. Wendy Panero, a geophysicist at Ohio State University, reminds us that water weathers rocks, rendering them more fluid, which "makes plate tectonics possible." And plate tectonics make the planet habitable for life by cycling heat, water, and chemicals across the sphere, keeping the Earth's climate relatively stable over eons of time. Columbia University geophysicist Donna Shillington sums up the new findings, which are changing the very ways geophysicists perceive the planet, noting that "water is as crucial to the workings of the Earth's interior as it is to Earth's surface processes."[42]

CHAPTER 2

The Earth Be Dammed

The Dawn of Hydraulic Civilization

The profound change in our species' worldview from thinking of our planet as *terra firma* to realizing that we inhabit Planet Aqua requires nothing less than reconceptualizing the way we categorize the fundamentals of our existence. Is our species "grounded" in the earth or do we exist as an extension of the waters? This means going back into our history and collective memories and examining the myths, legends, and stories we have told ourselves and passed on over the millennia that define who we thought we were, how we imagined life, and the way we experienced the planet we live on.

A running backstory has accompanied the many accounts of the origins of life and our species' cultural and social development over time contrasting an Earth-centric focus with an aqua-centric orientation. A careful exploration of our evolving consciousness inevitably leads to the great ongoing struggle to define our natural habitat. The question of whether we are of the clay of the Earth or of the waters of the ocean is central to our rethinking of the human narrative at this watershed moment in history. The increasing emphasis on how to survive on a warming planet that is resulting in the wild gyration of the hydrosphere is bringing to the surface a divided loyalty between securing the Earth and freeing the waters.

In fact, both need to be addressed but with the caveat that the securing of the Earth under our feet will depend on our recognizing and adapting to the freeing up of the hydrosphere that is taking life into uncharted waters. This will require reimagining what it means to live on a water planet in a tiny solar system nestled alongside seven other planets in the Milky Way galaxy. Leading the list of priorities will be exploring the ontology of water across history, focusing on the various dimensions of our intimate, but little thought about, relationship to the hydrosphere.

That relationship begins with understanding how our species came to "domesticate" the ebbs and flows of living on a water planet. It all started 11,000 years ago at the end of the last Ice Age and the dawn of a mild climate, which geologists called the Holocene, when our forebears, for the most part, left their nomadic life behind and began to settle into a sedentary existence, domesticating plants and wild animals and giving birth to agriculture and pastoralization – the rudiments of what we might generously call economic life. At the core of economic life is the fashioning of complex human infrastructures in which the hydrosphere comes to play a central role.

The Making of a Social Organism

Every great shift in the way our species interacts with the natural world since the dawn of civilization is traceable to epochal infrastructure revolutions, in which the hydrosphere is the first mover. Although most historians regard infrastructure as mere scaffolding to bind large numbers of people together in collective life, it plays a far more important role. New infrastructure paradigms bring together four components that are indispensable for maintaining a collective social existence: new forms of communication, new sources of energy and power, new modes of transportation and logistics, and most importantly new ways of managing the hydrosphere. When these four technical advances emerge and congeal in a seamless dynamic, they change the way a people communicate, power, and move their day-to-day economic life and establish their social and political norms.

As I described in a previous book, *The Age of Resilience: Reimagining Existence on a Rewilding Earth*, infrastructure revolutions are similar to what every organism requires to live: a way to communicate; a source of energy; mobility or motility to maneuver in the environment; and sufficient water to sustain life. Human infrastructure revolutions are a technological prosthesis that allows large numbers of people to interact in increasingly complex social, economic, and political settings, performing differentiated roles in what are defined as large-scale "social organisms."

Then too, every organism requires a semi-permeable membrane – for example, a skin or shell – to manage the relationship between its internal life and the external world, in order to survive and flourish. Our buildings and other structures act as semi-permeable membranes, enabling our species to survive the elements, store food and energies we need in order to sustain our physical well-being, provide safe habitats to produce and consume the goods and services necessary to maintain our existence, as well as serve as a congregating place to raise families and engage in social life.

Infrastructure revolutions also change a society's temporal/spatial orientation as well as the nature of social life, economic activity, and forms of governance conditioned by the opportunities and restraints that accompany the new more differentiated collective patterns of life made possible by the infrastructures.

In the nineteenth century, steam-powered printing, the telegraph, abundant coal, locomotives on national rail systems, and modern plumbing and sewer systems meshed with dense urban building stocks nestled around railway stations in an interactive infrastructure to communicate, power, and move society, giving birth to the First Industrial Revolution, accompanied by the construction of urban habitats, capitalist economies, and national markets overseen by nation-state governments.

In the twentieth century, the telephone, centralized electricity, radio and television, cheap oil, and internal combustion transport on continental road systems, inland waterways, oceans and air corridors, along with the engineering of artificial reservoirs and mega hydrodams that provided additional electricity and water for growing urban, suburban, and rural populations and intensive irrigation of agriculture, all converged. The Second Industrial Revolution gave rise to globalization and global governing institutions.

Today, we are in the early stages of a Third Industrial Revolution. The communication internet is merging with a digitized electricity internet powered by solar and wind electricity. Local and national businesses, homeowners, neighborhood associations, farmers and ranchers, civil society organizations, and government agencies are generating solar and wind electricity to power their operations. Surplus green electricity, in turn, is being sold back to an increasingly integrated, digitized, and soon transcontinental electricity internet using data, analytics, and algorithms to share renewable electricity as we currently share news, knowledge, and entertainment on the communications internet.

Now these two digitized internets are converging with a mobility and logistics internet made up of electric and fuel-cell vehicles powered by solar and wind-generated electricity from the electricity internet. In the coming decades, these modes of transport will be increasingly semi-autonomous and even autonomous on road, rail, water, and in air corridors, and managed by big data, analytics, and algorithms as we do with the electricity and communications internets.

These three internets are just now converging with a fourth digitized water internet, comprised of tens of thousands and soon hundreds of millions of distributed as well as decentralized smart cisterns and other water catchment systems, harvesting rainwater where it falls in the neighborhoods and communities where people live and work, and storing it in aquifers and water microgrids. Then, using big data, analytics, and algorithms, water is delivered across smart pipe systems, providing fresh water for drinking, bathing, cleaning, industrial uses, and irrigation of agricultural fields, while also recycling wastewater back to aquifers for repurification and reuse.

These four internets will increasingly share a continuous flow of data and analytics, creating algorithms that synchronize communication, the generation, storage, and distribution of green electricity, the movement of zero-emission autonomous transport across regions, continents, and global time zones, and the harvesting and purification of rainwater for human consumption, use by industry, and for the irrigation of farmland.

This ensemble of internets will also be continuously fed data from Internet of Things (IoT) sensors which monitor activity in real time, from ecosystems, lakes, rivers, and streams, agricultural fields, warehouses, road systems, factory production lines, and especially from the residential and commercial building stock, allowing humans to more effectively navigate day-to-day economic activity and social life.

The great infrastructure paradigm transformations represent the rise and fall of urban hydraulic civilizations across history. However, it's strange that more attention is traditionally placed on the economy, governance, and religious rituals that go along with the various stages of urbanization rather than focusing on the impact that sequestering the waters have had on making complex urban life possible. Nor have historians and anthropologists devoted much attention to how the rise and fall of urban hydraulic civilizations has for more than six millennia undermined the ecology of large swaths of the Earth's landmasses – transforming the waters of our planet from a "life source" to a "resource," leading to periodic thermodynamic collapse, the abandonment of large-scale human settlements, and the dispersion and mass migration of populations.

The hydraulic empires in Mesopotamia, Egypt, the Indus Valley of India, the Yellow River in China, and later the Roman Empire represented a new kind of human agency. Our ancestors began introducing highly complex technologies designed to harness and redeploy the waters – domesticating them to fit the exclusive temporal, spatial, and societal priorities of our species, where previously our species, like all others, had always adapted to the hydrosphere's temporal and spatial flows that accompany the daily rotation of the planet and its annual journey around the sun.

Other hydraulic civilizations emerged in Crete, Greece, the Khmer Empire in Southeast Asia, the Mayan civilization in Mexico and Central America, and the Inca Empire that spanned much of western South America, and each went hand-in-hand with the emergence of dense urban milieus.[1]

All of these civilizations sequestered their waters and erected complex hydraulic systems to provide a dependable and manipulatable supply of water for human consumption and to irrigate land and raise cereal grains.[2]

The first of the urban hydraulic societies was established by the Sumerians along the Tigris and Euphrates rivers, in Mesopotamia – a region that includes parts of modern-day Turkey, Iraq, Syria, and Iran. Thousands of laborers had to be indentured to build and maintain the dams, canals, and dikes. Specialized craft skills had to be developed to build the edifices and organize the production, storage, and distribution of grain. Architects, engineers, miners, metallurgists, bookkeepers, and the like made up the first specialized labor force in history. The Sumerians built magnificent city-states at Lagash, Nippur, Ur, Uruk, and Eridu and erected monumental temple structures to honor their gods.[3]

Most important, cuneiform, the first form of writing, was invented to manage the entire system. Everywhere that large-scale complex hydraulic civilizations were created – the Mediterranean, India, and

China – to grow cereal plants, human beings independently invented some form of writing to organize the production, storage, and logistics operations.

The first written script in Sumer dates to 3500 BCE.[4] Writing was used not only to organize commerce and trade relations but also as an administrative tool for managing government bureaucracies and religious practices and as an artistic medium for literary works. Clay bricks, parchment, papyrus, and wax tablets were all used as writing surfaces. Goose quills – called pen knives, and brushes – were used as styli, and inks from various vegetable sources were used to inscribe. The early writing was in the form of pictographic signs etched in clay. Phonetization followed suit. Special schools called "tablet-houses" were set up to teach scribes. Limited literacy ensured that skilled craftsmen, merchants, government officials, and palace priests could communicate among themselves. Later, the tablet-houses metamorphosized into schools, which became the center for learning in Sumer.[5] While it is estimated that less than one percent of the population in the Sumerian city-states was literate, the social, economic, and political impact of living in the first even partially literate societies was immense and far-reaching.[6] School courses even included the study of mathematics, astronomy, magic, and philosophy.[7]

Writing led to the codification and public proclamation of laws, putting the administration of justice onto a more systematic footing. When the Akkadians, who were Semitic-speaking, conquered the Sumerians and consolidated the formerly semi-independent walled cities into a single empire, they further refined the justice system. Bringing together people of diverse cultural and linguistic backgrounds required a code of laws that would apply equally to all and ensure justice for every subject. The Hammurabi Code – named after the ruler Hammurabi – was one of the first in history to ensure some limited rights of individuals, especially with regard to acquiring, holding, and inheriting private property. The Akkadians, who later founded the Babylonian empire, converted the Sumerian writing system to one that better suited their own language.[8] The Babylonian empire deserves a special place in the history of consciousness, for it was there that the "emergence of the individual in a literate society . . . occurs for the first time."[9]

By objectifying and codifying laws and ensuring a degree of individual legal rights, the justice codes had an effect on the development of individual self-consciousness. Establishing a common frame of reference for the administration of justice gave people an objective way to judge others' social behavior in relation to their own. The Hammurabi Code meshed the varied experiences of injustice and the prescribed remedies to address wrongdoings across many different cultures and abstracted them into more generalized categories. This forced a certain amount of introspection and interpretation on the part of the observer, to try to understand how his or her own experiences with injustice squared or were at odds with the prevailing norms, which were a composite of the experiences of disparate populations that crossed a number of tribal boundaries. By contrast, in a single tribal culture, the taboos were clear and unambiguous and required little reflection and interpretation. One didn't have to think about how one felt. One only had to conform his or her behavior to what the ancestors said was appropriate to the situation. Formulaic oral cultures called for formulaic emotional and behavioral responses, while literate cultures required individual emotional and behavioral responses geared to the unique circumstances of each novel situation, measured against the abstract norms established in law.

An abstract legal code undermined the traditional tribal authority by creating a set of laws that applied equally to all and required individual compliance, thus partially severing each male and female from their former collective tribal affiliation. The Hammurabi Code formally recognized, in a small way, the individual self as a separate being, for the first time in history.

The quantitative leap in cereal-grain production in Sumer vastly increased population, giving rise to what we would recognize as urban milieus – communities of upward of tens of thousands of inhabitants. Some of the production centers – the flour mills, ovens, and workshops – employed more than a thousand people, marking the introduction of the first urban workforces in history.[10]

Transforming entire river valleys into gigantic production facilities required a new form of highly centralized political control. The great hydraulic civilizations gave rise to the first government bureaucracies and autonomous rulers who oversaw every aspect of the administrative and productive life of the region – including mobilizing thousands of farmers during the seasons, cleaning canals, transporting, storing, and distributing grain, managing commerce with nearby regions, taxing the people, and maintaining armies to defend the borders. These early hydraulic civilizations also included thousands of specialized laborers, from palace priests and scribes to skilled craftsmen and servants. While it is estimated that less than ten percent of the populace – a fraction of the overall population – was engaged in specialized skills and callings, they represent the first inklings of civilized life.[11]

The price that human beings paid for civilizing the species was a mixed one. On the one hand, people were subjected to ruthless regimentation and rigid control. Every aspect of one's life was orchestrated by powerful bureaucracies under the tutelage of rulers who enjoyed absolute power. On the other hand, the creation of specialized craft skills and labor tasks and the invention of limited forms of personal property, money exchanges and wages, yanked individuals away from the collective "we" of prehistory, creating the first partial "selves" and a whiff of independence.

It was the Sumerian merchants, however, who gained the most in the development of hydraulic civilizations, enjoying far more independence and freedom than any of the skilled craftspeople. While they were expected to do the bidding of the rulers, they were also allowed to trade on their own accord. The merchants in Sumer became the first large-scale private entrepreneurs in history. Many of them amassed great fortunes.[12] The detached individual, unique and differentiated from the collectivity, breathed their first faint air of selfhood at Sumer.

All the major hydraulic civilizations built elaborate road systems and waterways to transport workforces and animals and exchange grain and other goods in commerce and trade. Kings' highways, which were used to maintain communications across the kingdoms, were among the other great inventions of the hydraulic civilizations, along with donkey and oxen-driven wheeled chariots and sailing ships. The royal roads of Babylonia, Assyria, and Persia inspired the Greeks and, later, the Romans, who copied and expanded their reach. Royal roads crossed all of India. A massive network of royal highways was also constructed in China with the advent of the empire in 221 BCE.[13] Royal highways made urban life possible.

The Sumerians, and every other hydraulic civilization since, developed rich urban cultures. Royal highways encouraged migration and facilitated city life. The urban centers became magnets for the mixing of cultures. The dense living arrangements invited cross-cultural exchange and the beginning of a cosmopolitan attitude. While new exposures created conflict, they also opened up the door to experiencing people, who heretofore had been considered alien and other as "familiars."

Finding the similar in alien others strengthened and deepened empathic expression, universalizing it beyond blood relationships.

The confluence of partial selfhood and day-to-day interactions with unrelated others, previously regarded as alien, was a breakthrough moment in human history. Exposure to individuals who were not part of their collective kinship group sharpened each person's own sense of individuality, if faintly. Granted, urban life can lead to isolation and aloneness. However, it also gives rise to a unique self, able to identify with other unique selves by way of empathic extension. Partially weaned from the collectivity, one begins the process of reconnecting to others, this time as distinct beings, furthering one's own sense of selfhood. The awakening of a universal sense of empathy for other human beings stirred slightly in the early hydraulic civilizations. It was enough, however, to mark the beginning of a new phase in the human journey.

The Invention of Economics

The emergence of hydraulic infrastructure is a transformational moment in the history of our species. *Homo faber*, the toolmaker, became an actor of far greater consequence by literally domesticating the whole of the hydrosphere of the planet, making it an instrument to serve the needs and aspirations of the human family. The bending of the Earth's hydrosphere to meet the exclusive will of our species has led to a huge increase in surplus food along with fewer hands needed in the fields, freeing up populations to migrate to more dense urban areas and introduce a new phenomenon called economic life. Urbanization, in turn, gave rise to what we call civilization. In case we think this is just a page from history, the whole of the twentieth century and the first two decades of the twenty-first century have seen the greatest expansion of hydraulic infrastructure in the past six thousand years, with the erection of a record number of hydroelectric dams and artificial reservoirs, pipes, and pumping systems, directed solely for the use of the human family. In 2007, the United Nations reported that a majority of human beings on Earth were now living in dense urban communities.[14] As of 2014, more than 400 cities around the world boasted populations of between one and 37 million people, made possible by intricate hydraulic infrastructure.[15]

While the great hydraulic civilizations and accompanying urban life appeared roughly around the same time in the Mediterranean, India, and China, and later in Southeast Asia and parts of Mexico, Central America, and western South America, anthropologists wondered why not elsewhere? A milestone study published in the *Journal of Political Economy* in 2021 by researchers from the University of Warwick, Reichman University, the Hebrew University of Jerusalem, Pompeu Fabra University, and the Barcelona School of Economics suggests an intriguing answer that forces a rethink of conventional theory about the origins of economic life and the accompanying creation of hierarchies that are the main stays of urban based hydraulic civilizations.[16]

The conventional wisdom goes something along these lines. The adoption of farming in the long expanse of the Neolithic period was a trial-and-error process, during which successive generations learned how to increase the generativity of land leading to incremental rises in productivity. These surpluses increased the human population, along with reducing the number of people needed in the fields, allowing redundant labor to migrate to urban areas, leading to the social stratification of life, the emergence of hierarchies, and the formation of states. If this is so, how then to account for the fact that agricultural regions in the tropics, with a similar history of trial-and-error, and an incremental increase in agricultural productivity, never "progressed" to highly structured urban life, with its complex hierarchies and state formation like the hydraulic civilizations. The researchers argue that "it was not an increase in food production that led to complex hierarchies and states, but rather the transition to reliance on 'appropriable' cereal grains that facilitate taxation by the emerging elite."[17]

Why cereal grains? Because, unlike root and tuber plants that decay quickly after harvesting and cannot be stored, accumulated, and distributed over time, cereal grains can be stored for long periods of time and distributed to the populace as compensation for indentured labor, or as a means of collecting taxes from the public that can later be used to pay others for services rendered, all of which makes cereal grains so attractive in the formation of social and economic hierarchies and government bureaucracies.

Surplus grain, arguably the first source of accumulated wealth in the form of capital overseen by governing elites and administered by bureaucracies, came with an equally compelling downside – banditry. Fending off would-be thieves and marauders from stealing the grain required policing authorities and the creation of the first militaries to protect the new form of stored wealth.

The researchers studied 151 countries between the years 1000 and 1950, and found that among the eleven principal cereals, only maize was available in the New World, whereas of the four main roots and tubers three were present in the New World before 1500 – cassava, white potatoes, and sweet potatoes. By contrast, in the Old World, only yam was present among the main roots and tubers.[18] Yet, at the same time, all cereals but maize were available, explaining why, in the New World, only Mexico and Peru created bureaucratically ministered state regimes. Regarding the premise that cereals "played a crucial role in state formation," the researchers found that "in farming societies that rely on roots and tubers, hierarchical complexity never exceeded the level that anthropologists define as chiefdoms, while all agriculture-based large states that we know of relied on cereals."[19]

But here's what their study left out. Cereals require substantially more water than roots and tubers. A recent study comparing the water footprint of roots and tubers to cereals in the journal *Environment, Development, and Sustainability*, confirms many earlier studies that "the water footprint of roots and tubers is substantially smaller than that of cereals." What this means is that the regions of the world that grew cereal grains required substantially more water for irrigation and that is why civilizations dependent on cereals as a mainstay of their diet – throughout history – have built complex hydraulic infrastructures to sequester massive volumes of water to irrigate their fields and grow their crops.[20]

Nowhere was the connection between hydraulic infrastructure, the scaling of surplus grain, and the rise and fall of empires more in evidence than the Roman Empire. The Roman Empire excelled in the design and deployment of a vast hydraulic infrastructure across much of Europe and the Mediterranean. Its aqueducts harnessed water for drinking, hygiene, and pleasure and, importantly, for irrigation of farmlands stretching from Spain and France to the West and Egypt, Syria, and Jordan to the East.[21]

The cultivation of grains at-scale across a complex hydraulic infrastructure was critical to securing the stability of the empire. The governments' very viability teetered, in no small part, on the dispersing of free bread weekly to its hundreds of thousands of citizens and slave workforce. The grain, mostly durum wheat, was grown across the empire and transported via sea and terrestrial routes to Rome and other provinces. Egypt was a principal grain hub for Rome and the provinces, as were the regions that today encompass Libya, Tunisia, Algeria, and Morocco.

Commenting on the overwhelming importance of providing free grain and bread to Roman citizens, the satirical poet, Juvenal, mused that a waning empire, whose former glory on the battlefields of Europe, Asia, and Africa, pillaging and colonizing parts of three continents, had lost currency, was only able to maintain its weak hold by offering up "bread and circuses" to the people. This free-bread arrangement eventually collapsed because of the deterioration of the empire's farmlands, weakening the empire's hold on its citizenry and making this once-great power vulnerable to marauding invaders and, eventually, the invasion by the Vandals and the sack of Rome in 455 CE, and the collapse of the empire.

The hydraulic empires represented a new kind of human agency, exercised collectively in dense urban environments. Before their appearance, rudimentary urban life in the form of villages and incipient urban enclaves was common. Jericho is regarded by many anthropologists and historians as the first urban commune – a city of a few thousand inhabitants at best and, even then, somewhat transient, as the population ebbed and flowed with the seasons. Communal life was organized around polytheistic religious practices and idolatry in the form of fertility cults and occult practices. Later on, the Israelites conquered the city, establishing the first enclave of the Hebrews upon entering the "Promised Land."

Most of the early primitive urban milieus acted more or less as trading posts visited by itinerant bands. They were also protectorates – walled off edifices – protecting local populations from marauders. But they lacked identifying markers of what we call urban existence. There was very little differentiation of skills and only sketchy governance.

These early forms of natal social existence were almost always sited near lakes, rivers, and well water adequate to sustain small populations. Still, these habitats took our species a step beyond villages and closer to urban life in the form of "commons governance" and continue to exist in scattered locations to the present day, although now embedded within sovereign national borders.

But urban life, as we've come to know it, is all about large numbers of unrelated individuals in the tens of thousands, hundreds of thousands, or even millions, living together in complex environments replete with a myriad of differentiated skills in what early on became known as arts, crafts, and trades and now companies, industries, and sectors underwritten by codified laws regulating economic activity and exchange. These are the essential components of what we call urban life. And this configuration – a social organism – only emerged with the rise of hydraulic civilization in the millennia before Christ, and ever since expanding and collapsing over history and becoming the dominant way our species organizes its governance and economic and social life.

A debate of late has emerged regarding the pattern of governance that has accompanied the rise of urban life. In their book *The Dawn of Everything*, David Graeber and David Wengrow make the case that early urban life was more in the form of commons governance, with only a hint of political hierarchy and bureaucratic oversight. The authors point out that these early urban habitats were in many instances remarkably democratic in spirit, painting a picture of economic and political life that more resembles the rudiments of what historians today define as "civil society," all of which makes sense. Commons governance in the form of reciprocal exchanges of goods and services and a collective stewardship of a shared natural environment is the bedrock of social life – the floor in the buildup of urban civilization through history.[22]

What's left out of the Graeber and Wengrow analysis is that in various regions of the world where our forebears laid down roots to establish a collective urban life, many of these early commons forms of governance reached a stage of population density that required some form of sequestration of the planetary hydrosphere to provide sufficient water and food for their populations. Yet, strangely missing from the authors' otherwise astute analysis of how early urban life came to be, with its emphasis on commons governance and what may be regarded as a nascent civil society and a fledgling democratic approach to governance, is what transpired once the population exceeded the bounds of its available food sources and provisioning of water. This is the point at which in some regions of the world urban commons morphed into a complex social organism of the kind we today think of as urban civilization.

The German American scholar Karl Wittfogel gave a name to this exceptional change in the way our species lives on Earth with the publication of his book *Oriental Despotism* in 1957.[23] He called this great transformation "the hydraulic civilization" – the mass extraction, sequestration, propertization, commercialization, and consumption of the planet's great rivers, lakes, wetlands, floodplains, and well water for the utilitarian use of our species. Wittfogel's analysis was not without shortcomings; some the result of ideological blinders and cultural biases, and others because of improperly identifying a number of the key features and components that gave rise to urban hydraulic civilization. Nonetheless, he gave a name to this developmental process, which has been with us for more than six millennia. The urban hydraulic civilization is what binds the human race together, with only a tiny number of our species still living on the periphery or outside the bubble.

Wittfogel's seriously flawed but nonetheless provocative identification of large-scale hydraulic infrastructure as the core of what we have come to know as urban civilization – a six-millennia journey that has now taken us to the edge of extinction – is indisputable. Commons governance in the form of an evolving civil society continues to be an important component and, at times, opponent of hydraulic civilizations. Nothing more and nothing less.

The late Robert Leonard Carneiro, an American anthropologist and former curator of the American Museum of Natural History, attempted to make peace between the two warring factions on the question of the underlying conditions that gave rise to state formations. He wrote, "This is not to say, of course, that large-scale irrigation, where it occurred, did not contribute significantly to increasing the power and scope of the state. It unquestionably did. To the extent that Wittfogel limits himself to this contention, I have no quarrel with him whatever. However, the point at issue is not how the state increased its power but how it arose in the first place."[24]

Granted . . . but this is a bit of red herring. Obviously an urban foundation of sorts and the presence of a budding commons governance and civil society is a quid pro quo for developing a rudimentary state, without which the rise of a complex urban hydraulic civilization would be an impossibility. But, left unsaid by Graeber and Wengrow and others is how do we account for the fact that the great urban civilizations of antiquity that we continually revisit rest on harnessing of the great rivers of the world in some form or another to survive and flourish.

This doesn't suggest that the evolution of urban hydraulic civilizations have all followed a similar growth pattern. Consider, for example, the route taken in feudal Europe after the fall of the Roman Empire. While most of us are taught that Europe regressed into what historians call the Dark Ages, nothing could be further from the truth.

Although the great Roman aqueducts across Europe were abandoned and, for the most part, left to rot in subsequent centuries, a fledgling new approach to sequestering the hydrosphere began to take hold across the continent as local artisans installed water mills on European rivers. By the eleventh century, 5,600 water mills, stretching across 3,000 communities in England alone were used to mill grain.[25] Water mills were omnipresent across Europe, and used not only for milling grain but also for laundering, tanning, sawing, crushing olives and ores, operating bellows for blast furnaces, reducing pigments for paint or pulp, and polishing weapons.

These water mills were ubiquitous in urban communes along their riverfronts. To get a feel for their importance in revving up a proto-industrial revolution, in France alone there were 20,000 water mills in operation by the eleventh century, or a mill for every 250 people.[26] If calling this phenomenon "industrial" seems a bit out of place, consider that in the water mills a single worker at the helm replaced the labor of twenty workers – an impressive leap in productivity by any measure. The combined hydraulic power of these thousands of water mills was equivalent to the labor power of 25% of the adult population of Europe.[27] By the 1790s, there were 500,000 water mills across Europe, sequestering the hydraulic flow of water across every major river, providing 2,250,000 horsepower.[28] Princeton University historian Lynn White summed up the historic significance of sequestering the waters of Europe for industrial purposes well before fossil fuel energy came onto the field remarking that as early as "the latter part of the fifteenth century, Europe was equipped not only with sources of power far more diversified than those known to any previous culture, but also with an arsenal of technical means of grasping, guiding, and utilizing such energies, which was immeasurably more varied and skillful than any people of the past had possessed . . ."[29]

While the medieval water mills were decentralized and overseen locally, the lessons learned in managing the rivers of Europe were put to work in the late nineteenth and twentieth centuries in constructing massive hydroelectric power dams and artificial aquifers and accompanying water infrastructure. Britain led the way in 1878 with the first proto hydroelectric power experiment. Soon thereafter, America and countries around the world introduced hydroelectric power dams to serve a global industrial economy. And, alongside the sequestration of the waters across Europe, America, Asia, and the rest of the world, governments established centralized bureaucracies replete with codes, regulations, and standards, to manage what became the pinnacle of hydraulic civilization.

The point of all of this is that the rise and fall and rise again of urban hydraulic civilizations across six millennia of history is a defining element of the human journey, although rarely recognized. More decentralized and democratized commons governance of our planetary hydrosphere has increasingly become a footnote to history.

What is particularly striking in the accounts of the emergence of urban-based hydraulic civilizations is that little has changed in the past six thousand years in how workers have been employed, compensated, and taxed. By the time of the third dynasty of Ur in ancient Sumer large workforces were a way of life, and included a raft of farmers, urban laborers, skilled craftsmen, merchants, bureaucrats, priests, servants and menial labor and a long list of other forms of work.

The Sumerians were likely the first urban civilization to envision labor as an abstract concept – a quantifiable measurement of a person's work by classification and skill, and whose compensation was measured in workdays performed and output produced. As early as 2400 BCE, the Sumerian word *á* meant "arm, strength, power, and physical exertion" signifying that work was already seen as something that could be logged and renumerated – representing the earliest record of what would later become "labor" and, in the modern era, "employment."[30]

Harvard University's Piotr Steinkeller, a professor of Assyriology, suggests that the "ability to count labor in abstract 'workdays' (or 'man-days') was a conceptual breakthrough in the history of accounting and administration, since it allowed the conversion of any form of productive human activity into a set of numbers, therefore opening up completely new managerial possibilities, particularly in the area of economic planning."[31] Steinkeller notes that the word *á* eventually expanded to encompass "wage, hire, rent," showing that "labor not only was a function of time but that it also had a measurable monetary value, which could be expressed in terms of silver or grain." Economists have long believed that the invention of double-entry bookkeeping was the brainchild of merchants in Venice in the late fifteenth century, laying the groundwork for the rise of modern capitalism, when in fact the rudiments of modern accounting date back to the first hydraulic civilization in Sumer, 2,000 years before Christ.[32]

It's also widely assumed that the first hydraulic civilizations relied primarily on slave labor to build out, manage, and repair the hydraulic infrastructure and serve as labor in the fields and elsewhere. While slavery existed in the first hydraulic empire in Mesopotamia, slaves were small in number and generally assigned to domestic work as servants in wealthy households or as craftsmen. The majority were debt-slaves who, upon repayment of their debts, would be freed.

The primary source of labor in Mesopotamia was forced labor or *corvée*. This was work "required" by the state, generally for several months of the year, and it encompassed the entire population. Steinkeller notes that few individuals and families were exempt. "Craftsmen, shepherds, agricultural personnel, gardeners, foresters, merchants, and various types of administrative and cultic officials, as well as members of local elites, such as the provincial governors and their kinsmen" were expected to devote several months of the year to *corvée* work – although it's likely that the very wealthy and government elites did not enlist but found substitutes to do the work.[33]

While the populace was technically free, everyone was required to put in their time in maintaining the hydraulic irrigation system, harvesting the crops and building and repairing palaces, temples, city walls, and other public structures. Military service was also among the *corvée* public works obligations. These workforces were rewarded by the state with access to agricultural land and irrigation waters. *Corvée* was a form of imposed communal work, which everyone was expected to carry out as a service to their community and the state, in return for receiving some of the collective benefits that came with their participation.

There were also other lesser work categories, including menial labor attached to temple households and other state institutions. Hired labor and contract labor were two other categories of work, each with their own constraints and rewards. What's interesting about *corvée* is that it was looked on as a communal endeavor and collective responsibility – a sort of "commons" form of "public" participation, binding the populace together as a social organism – the first stirring of what would evolve into the civic square and the rudiments of civil society. What we are left with is this: the evolution of dense urban life and civilization are one and the same. One does not exist without the other, and both together require some form of hydraulic infrastructure to function.

What's so extraordinary is the dearth of knowledge about the role that hydraulic infrastructure plays in the rise of urban culture and the emergence of civilization. And we're not talking about long ago in ancient times. The fact is that hydraulic infrastructure is the irreducible scaffolding upon which all of urban life depends and the indispensable platform upon which civilization rests. Yet, in all of the weighty discussion and philosophical and anthropological debate about how urban life came to be, barely a passing reference is given to the underlying hydraulic infrastructure that makes large-scale urbanization viable.

Middle-class urban dwellers in the wealthy industrial nations rarely give thought to where water comes from, how it's purified and managed, and where it goes after we've used it. The urban poor, however, are often acutely aware of its importance, especially when it's contaminated with toxic chemicals or when the aged pipes are laden with lead. In developing countries, where water is at a premium and rarely safe to drink, or unavailable, the poor are continuously alert to how the water system is functioning. Rural residents, and particularly farmers, are well versed in the intricacies of hydraulic flows and the workings and failings of the hydraulic infrastructure, as their very livelihood depends on irrigation.

It goes unsaid that several billion human beings spend much of their waking hours worrying about the availability and purity of water and sense that something is awry. Sometimes they blame God and/or the climate for withholding the water, and at other times claim, with some truth, that nefarious forces are at work, often among urban elites gaming the flow and availability of water to their advantage. But it's more than that. Controlling the waters is the ultimate playing out of "man's" dominion over Planet Aqua. It's a conscious choice, first made by our species several thousand years ago in disparate regions of the world. Capturing the waters flowing across powerful rivers that stretch for hundreds of miles, and then holding them at bay and administering them methodically, to optimize agricultural yields and store grain surpluses and then coordinate the timing of food distribution, and to whom and in what quantity is what defines the hydraulic civilization. The sheer scale of this human-designed and deployed infrastructure is without parallel. This exercise of raw power over the planet's hydrosphere is the foundational infrastructure upon which all of economic life depends.

Hydraulic infrastructure is by its very nature a centralized construct, hierarchical in design, bureaucratic in deployment, and coercive in execution. While human beings design, engineer, deploy, and manage hydraulic infrastructure, it's the infrastructure that determines the kind of society that's compatible with the workings of that infrastructure. Hydraulic infrastructures change the temporal and spatial orientation of society, and condition the kinds of governance that are compatible with their design and operation.

Even a society's habitats, child-rearing practices, approaches to learning, and relationship to nature are conditioned by the limits imposed by adapting the waters to our species rather than adapting our species to the waters. That doesn't mean a rigid conditionality but only that the infrastructure sets the boundaries that are compatible with the infrastructure. The options can be varied but not be at odds with the way that the hydraulic infrastructure is constructed and deployed.

The point is, worldviews don't create hydraulic infrastructures. It's hydraulic infrastructures that create worldviews. Hard to accept, perhaps, because we have come to believe that human agency determines how infrastructure is used. Try to imagine for a moment the incredible hubris in believing that our single species, regardless of our outsized brain, could have ever believed that we could hold dominion over the waters of the Earth. Yet, that's exactly what we set out to do in scattered regions around the world, some six millennia ago. And the result is that after six millennia of domesticating waters, the entropic impact on the planet and the rest of life on Earth is staggering to behold. Although *Homo sapiens* makes up less than one percent of the earth's total biomass, by 2005 we were using 25% of the net primary production from photosynthesis, and current trends project that we might use as much as 44% by 2050, leaving only 56% of the net primary production for the rest of the life on the planet ("net primary productivity is the net difference between photosynthesis and respiration . . . it's how much carbon dioxide plants take in during photosynthesis minus how much they release during respiration").[34] This extraordinary seizure and consumption of the Earth's storehouse lies with the sequestering of the waters and managing their deployment via hydraulic infrastructure.

Much of the human conflict over the history of hydraulic civilizations has been around access to the waters. We ought to remember, however, that for most of our time on the planet our species considered the waters to be an open commons shared by the human family, along with our fellow creatures. The commons approach to the hydrosphere continued to flourish in scattered places, even with the incursion of hydraulic civilization.

Although the sequestering of the hydrosphere precipitated a giant step forward in the storage of cereal grains and the distribution of surplus food, dramatically increased population and human longevity, freed up labor in the fields, and led to the emergence of urban life, the constraints imposed by the very nature of the hydraulic infrastructure has also been a principal catalyst of conflict and warfare among our species for millennia of history.

While our ancestors knew this early on and continuously engaged in conflict and open warfare over access to sequestered hydraulic waters, the question of whether the infrastructure itself ought to be the subject of concern has rarely been examined. Perhaps this is because the benefits of sequestering the waters to increase food production of cereal grains and create food surpluses, free labor to migrate to cities, improve living standards and increase the life expectancy of our human family was too powerful a tonic to question. Or, perhaps it's because hydraulic infrastructure became so pervasive in every aspect of life that the infrastructure came to be seen as an inextricable part of the landscape and a reflection of "the natural order of things." And that natural order came to be seen in the Western world as a gift from an Almighty God, granting our species dominion over the Earth and all living things.

Recently an awakening of sorts has occurred, with the warming of the planet brought on by a fossil fuel-based industrial era. The hydrological cycle is rewilding, freeing itself from the bonds of a hydraulic civilization. The coming of powerful atmospheric rivers, damaging floods, prolonged droughts, heatwaves and wildfires as well as hurricanes and typhoons has shaken scholars who are beginning to raise concerns about the very operating assumptions that underlie hydraulic infrastructure and questioning the fabric of what we call civilization.

In 2022, researchers from the Wageningen University and the University of Amsterdam in the Netherlands – a country that can claim to have a rich history of sequestering the waters and deploying a hydraulic infrastructure in Europe – published a study in the journal *Political Geography* entitled "(Re)Making Hydrosocial Territories." Their study reflects a brewing debate gathering momentum in rethinking how the sequestration of the waters and the evolution of hydraulic infrastructure act "to transform relations between space, people and materiality."[35]

The researchers studied hydraulic infrastructures on every continent with the premise that they are not really neutral scaffolding but rather embrace a worldview – what the researchers refer to as an "imaginary" – that plays an integral role in conditioning how we live our lives, including understanding our role in the economy, defining our place in society, determining our relationship to nature, and even how we affiliate with governance. The scholars begin with the obvious: the grand hydraulic infrastructures, including large dams, irrigation schemes, and hydropower plants that are all around us, and are viewed as a centerpiece of what we prize as "modernity" and "progress." They write that:

> Modernity in this context is often associated with key characteristics such as the belief in continued progress, the belief in planned social, ecological and technological futures, the centrality of science and technology in this planning process, and the need to control and domesticate nature.[36] Especially the last two aspects are intrinsically connected to hydraulic infrastructure as they have made it possible to enroll nature as an economic resource in intensifying and expanding modern production systems. At the base of these undertakings is a modern imaginary of nature as external to society, as disordered, savage and something to be controlled and put to productive use through advancing science and technology. Nature is thus imagined as an entity that awaits to be mastered and turned productive for societal benefit. Such imaginaries that envisage modernizing territorial transformations through infrastructure aim to dramatically alter the spatiality and materiality of landscapes, water flows and importantly also the social and political relations in these.[37]

Hydraulic infrastructures, however, didn't emerge out of the blue. Our forager hunter ancestors could not have conceived of such an idea.

They led a nomadic existence, following the seasons and adapting moment to moment to whatever nature provided. The receding of the last great Ice Age 11,000 years ago brought on a temperate, arid, and semi-arid climate in large parts of the world, conducive to sedentary life and experimentation with growing food and domesticating animals for food and fiber.

It's likely that the first forays in adapting nature to our species rather than our species to nature came with planting seeds, harvesting crops, and finding hardy and nutritious new varieties. It's not a big departure from there to developing a rudimentary belief in dominating nature – an idea that would grow in time and lay the path from stewardship to subjugation of the natural world. Sequestering and domesticating the waters to ensure a hearty harvest and surplus would have been a small leap but with epochal consequences that would come to alter our species' relationship to the natural world, culminating in the industrial hydraulic infrastructure that became a foundation of the Age of Progress.

Hydraulic infrastructure also sets the territorial boundaries of governance as well as designates who designs and manages the system, how the waters are allocated, to whom, and under what conditions. In short, the infrastructure determines power relations at every level of society and across the territory where the hydraulic infrastructure flows.

The authors of the study suggest that hydraulic infrastructure gives "answers to the moral questions of how to live and how to act ... in other words, water technology is 'moralized,' bearing its designer's class, gender, and cultural norms, and actively proliferating these moral and behavioral norms when the technology is applied." What often goes unnoticed is, "by redirecting water flows and changing landscapes, the ways people relate to and experience their environment change ..."[38]

Sequestering, domesticating, and redirecting the planet's hydrosphere to create a cornucopia for the human species is the ultimate utopian dream put to practice. And, like other utopias, its promises were never realized in the final analysis. A close look at the expectations and deliverables paints a dismal record of false hopes and calamitous consequences across history. Study after study has documented that irrigation systems have systematically underperformed over their "lifetime" in terms of increases in agricultural productivity. There are a host of reasons why. Here's a list of some of the particulars:

> Domestic water supply systems lose water and don't deliver the quantity and quality water that was projected. Hydro-powerplants rarely produce the promised electricity outputs. These faults and fractures come from the unpredictability of (a) nature (floods and droughts, soil erosion, and sedimentation, etc.), (b) infrastructure and its intrinsic properties in use (wear and tear), and (c) the social system that controls, manages and uses the infrastructure and related water flows.[39]

It's not that the shortcomings and failings of hydraulic infrastructures were a secret. Over six millennia of history, the rise and fall of hydraulic infrastructures has marked the life cycles of entire civilizations – each inevitably caught up in the stranglehold of a rising entropic bill and dramatic changes in the climate. Still, faith in the enterprise has continued to rebound, likely because of an unquestioned belief – especially in the Western world – in our species' special covenant with the Lord, granting human beings mastery over all of the Earth. Adapting nature to our species rather than the other way around is the signature of civilization and the very heart of a worldview we have come to internalize during much of the Holocene era. If there is an excuse to explain our folly, it is likely that the temporal climate proffered the illusion that we could sequester the hydrosphere and recondition the workings of Planet Aqua to do our own bidding without fear of reprisal. How wrong we were.

Strangely, the slight mentions of hydraulic civilization, usually as footnotes in anthropology and history seminars and rarely, if ever, a subject of public discussion, gloss over their significance, as if to suggest that they are relics of a history long buried and treated as archaeological discoveries . . . not so. Hydraulic civilization reached its peak in the past two centuries of a fossil fuel-based industrial era. It's the great undergirding that has taken us to the commanding heights as the dominant species on Earth and to the sixth extinction of life on the planet.

Drowning in Progress

Hydraulic civilization's apotheosis came with modernity. The Age of Progress is tautologically tied to the sequestering of the hydrosphere of the planet, although barely acknowledged. The discovery of the "new world" and the colonization of indigenous populations across every continent in the nineteenth and twentieth centuries rode alongside two pillars – military conquest and the "reeducation" and "re-politization" of native peoples. That re-politization came in the guise of building out and managing gigantic hydraulic infrastructures. Control of the waters meant control of entire populations. The new colonial overlords tutored a new class of native elites – mostly engineers educated abroad in the arts and sciences of hydraulic deployment and management. The establishment of water bureaucracies and governing statutes, codes, and regulations brought together colonial rulers and a new class of colonized bureaucrats to jointly manage hydraulic infrastructures in "developing countries." *Corvée* labor was recruited and more often indentured to build out and operate these modern hydraulic infrastructures.

The agricultural surpluses brought on by increasingly sophisticated hydraulic management were shipped back to home ports in Europe, the Americas, and elsewhere, providing relatively cheap produce and fiber at the expense of cheap foreign labor. Later on, and especially after World War II, the so called "third world" began struggling to free itself of colonial rule, while keeping the colonial infrastructure intact, only to upgrade it, further sequestering the waters to serve their domestic populations, while depleting their remaining ecological capital.

The latter half of the nineteenth century, the entirety of the twentieth century, and the opening decades of the twenty-first century were the high-water mark for hydraulic infrastructure deployment everywhere in the world, as nations vied to outperform each other, building ever-bigger dams, harnessing greater volumes of remaining river water, piping water over hundreds of miles, often across arid and semi-arid lands, and even deserts.

The magnitude of damming up major rivers of the world over the course of the last hundred years, and which continues to this very day, is impressive. There are currently 36,222 dams spread across major river basins in North America, South America, Europe, Africa, and Asia. Asia scores the highest number of dams completed with 10,138 infrastructures, and represents 28% of worldwide dam construction. North America represents 26% of the dams while South America accounts for 21% of the global dams built out in the past century. Asia and South America make up 50% and 20% respectively of global installed capacity while North America represents 9% and

Europe 18% of installed capacity. The dams are used for provisioning fresh water for human consumption, cleaning, and bathing, generating hydroelectricity for homes, offices, and industries, and irrigating agricultural fields.[40]

Although dam construction in the United States has slowed over the past several decades, mainly because all of the rivers have already been dammed, the developing countries in Africa, South America, and Asia are increasing dam construction.[41] America led the way in the last great push to dam all of the rivers of the world. Its greatest triumph was the rerouting of the Colorado River to flow into the artificially constructed Lake Mead and from there flow over the newly constructed Hoover Dam, providing electricity and water to 25 million people in the states of Arizona, Nevada, California, and parts of Mexico.[42]

Turning semi-arid and arid lands, and even deserts into fruit orchards and amber waves of grain, not to mention the erection of cities, suburbs, resorts, and golf courses, was never a good idea – still, a constant reminder of the hubris that accompanied the vision of a hydraulic utopia. Other countries everywhere followed along a similar line.

In the opening decade of the twentieth century, the great enthusiast of hydraulic civilization, William Ellsworth Smythe, eulogized the scientific conquest of the waters in a book entitled *The Conquest of Arid America*. He opined "irrigation is the foundation of truly scientific agriculture. Tilling the soil by dependence upon rainfall is, by comparison, like a stagecoach to a railroad, like a tallow dip to the electric light."[43]

All across the world, throughout the twentieth century, the clarion call among scientists, engineers, entrepreneurs, politicians, and business leaders was to tame the rivers, so that a "desert [could be turned] into gardens,"echoing the long-held belief that our species' dominion over nature could pave the way back to a second Garden of Eden, this time by way of a utopian dream of sequestering the waters.[44]

What's changed over the course of the past several decades is a revolt of the hydrosphere, caught up in a quickly warming climate, decimating ecosystems, destroying urban and rural communities, taking human lives and the lives of our fellow creatures and disabling the hydraulic infrastructure everywhere. The warming of the climate has awakened a younger generation to the folly of ever believing that our species could sequester and harness the entirety of the planetary hydrosphere to suit the commercial and political machinations of a beguiled humanity.

The awakening from our longstanding hydro-utopian nightmare began with the birth of the new science of ecology in Europe in the nineteenth century, followed by the conservation movement and the establishment in America of national parks and wildlife animal preserves in the early twentieth century and, later, the emergence of the environmental movement of the 1960s, as well as, two decades later, the green movements in Europe and America respectively, and now the meteoric rise of the Millennial and Gen Z generations, who see themselves as "an endangered species" alongside our fellow creatures on a dying planet.

It's the younger generation who are beginning to sleuth the particulars that led to an extinction event, and are coming to understand that the sequestering of the waters and the emergence of civilization are causally interrelated. What then? The Millennial and Gen Z generations are beginning to see the freeing of the hydrosphere as a revolutionary act and an act of redemption.

The hydrosphere is rewilding and searching for a new normal on a warming planet and in the process taking all the Earth for a wild ride. But, taking this moment forward, a new planet is striving to be reborn. How then to define the future way to live on a very different Earth, which is fast leaving what we call civilization behind? When we think of civilization, it's a concept easy to understand yet difficult to articulate. Civilization is a place and a way of life. When we conjure up civilization, it's always in relation to "the other," which has usually meant an alien nature, at least in the Western world. Nature is generally perceived as wild, untamed, precarious, and even savage and brutal, and a phenomenon to be pacified, tethered, modified, and consumed. At best, nature is viewed as a resource and, at worst, a dangerous and menacing foe. Civilization, on the other hand, is seen as a safe haven and domicile. It's where we domesticate and cooperate at scale to secure our collective well-being as a species. Civilization, then, comes to mean the predictable ordering of time and space, largely cordoned off from an unpredictable nature.

Civilization is also about advancing and perfecting human existence, a slippery term that usually takes on significance in relation to what it isn't, that is, "nature, red in tooth and claw."[45] That's not always been the case. The eighteenth-century French philosopher Jean-

Jacques Rousseau was of the mind that in the state of nature "human beings are good by nature but are rendered corrupt by society," a theme echoed by other philosophers, who became known as the Romantics.[46] Others sided with the philosopher Thomas Hobbes, who believed that human beings in the state of nature are biologically inclined to fight one another to secure the basics of life in a war of each against all, adding that life is nasty, brutish, and short in a state of nature. He argued that only by giving up a measure of one's freedom in nature and accepting the heavy hand of a strong governing power, backed up by strictly enforced laws and rules of conduct, accompanied by harsh punishments, could man be coerced to cooperate for the common good.

Immanuel Kant, argued that man is by nature a rational being, but ought not be led astray by emotions and feelings and the physicality of existence. His ideas sparked the Age of Reason and the notion that civilization is a rationalizing exercise and, in its pure state, detached from the environment, further separating human beings from the natural world. His thoughts generally prevailed, giving rise to what philosophers call the Enlightenment and, soon thereafter, the Age of Progress.

Progress is a French word and first appeared in print in Europe in the eighteenth century. The word was immortalized by the French philosopher Nicolas de Condorcet. In the dark days of the French Revolution in 1794 he penned a short essay that became the leitmotif of modernity and the most up to date understanding of the primary purpose served up by civilization. He writes, "no bounds have been fixed to the improvement of the human faculties . . . the perfectibility of man is absolutely indefinite . . . [The] progress of this perfectibility, henceforth above the control of every power that would impede it, has no other limit than the duration of the globe upon which nature has placed us."[47] In a way, the idea of progress is not all that new a concept, but a more recent take on the long-held belief in the Western world that the descendants of Adam and Eve are to rule over nature, sequester it, and utilize it to advance society and edge ever closer to a facsimile of the original Garden of Eden, and a welcoming back to Paradise.

All civilizations are wrapped up in utopian claims and, at the head of the pack, is the vanquishing of barbarism that was thought to characterize human existence in a state of nature. The political scientist, Anthony Pagden, of the University of California, Los Angeles, says that civilization "describes a state, social, political, cultural, aesthetic . . .

which is held to be the optimum condition for all mankind" . . . edging ever closer to a utopian vision.[48]

If we are looking for a widely accepted appraisal of the nature of civilization and its nemesis of an unbridled nature, John Stuart Mill hit the mark back in the nineteenth century in an essay that came to define both the colonial era and modernization, although with roots that wind back to the earliest reckonings of hydraulic empires and the emergence of civilization in the centuries before Christ. He writes:

> a savage tribe consists of a handful of individuals, wandering or thinly scattered over a vast tract of country: a dense population, therefore, dwelling in fixed habitations, and largely collected together in towns and villages, we term civilized. In savage life there is no commerce, no manufactures, no agriculture, or next to none; a country in the fruits of agriculture, commerce, and manufactures, we call civilized. In savage communities each person shifts for himself; except in war (and, even then, very imperfectly) we seldom see any joint operations carried on by the union of many; nor do savages find much pleasure in each other's society. Wherever, therefore, we find human beings acting together for common purposes in large bodies, and enjoying the pleasures of social intercourse, we term them civilized.[49]

The notion of characterizing all other people outside the realm of "civilization" as backwards and savage provided the colonial powers with the moral authority they needed to capture, subdue, and exploit native populations throughout the world, justifying their plunder and expropriation with the admonition that they were bringing civilization to the backward, downtrodden, unwashed masses.

The French historian François Pierre Guillaume Guizot zeroed in on how scholars have come to describe civilization, at least since the dawn of the Enlightenment – although, it could be argued that the principal assumptions date far back in time to the first hydraulic civilization in ancient Mesopotamia. He argued that "the first fact comprised in the word civilization . . . is the fact of progress, of development; it presents at once the idea of a people marching onward, not to change place, but to change its condition; of a people whose culture is conditioning itself and ameliorating itself. The idea of progress, of development, appears to me the fundamental idea contained in the word *civilization*."[50]

Largely missing in all of the scholarly discussion about the nature of civilization is what was the precipitating agent that weaned it into existence in disparate parts of the world around the same time in history? What tipped the scales back then from village life and small-scale agriculture and pastoralization to dense urban centers of tens of thousands of people and more? It was the development of hydraulic infrastructure that gave rise to civilization. Adapting the planet's hydrosphere to the exclusive use of our species marks a turning point in humanity's relationship to nature, severing our ties to the environment and beginning our long journey to distance our species from the Earth.

The hydraulic civilization reached new heights across the twentieth century as scientists, engineers, and water bureaucrats teamed up with governments and industry everywhere, each competing to outdo one another in a show of technological swagger unmatched in the history of public works. The rallying cry to enlist popular support was much the same everywhere . . . to promote a "scientific agriculture" and to "let the desert bloom," a shoutout to farming communities and a new generation of urban and suburban dwellers seeking the warmth of a semi-arid and arid climate. In the United States earlier generations had referred to the laboring class as "rednecks," a disparaging remark meant to identify unskilled and semiskilled labor, who must work outside in a beating-down sun by the "sweat of their brow." This slur was transformed by middle-class families who came to idealize the semi-arid and arid lands with a more leisurely good life, characterized by a perpetual suntan.

The conquest of nature became a universal battle cry. Fittingly, the United States inaugurated what would become a century of hydraulic overshoot, with the erection of the majestic Hoover Dam – originally named the Boulder Dam – crowned in popular lore as one of the seven industrial wonders of the world. The sheer size of the dam mesmerized the public and positioned the U.S. as the lead hydraulic power of the twentieth century. President Franklin Delano Roosevelt inaugurated the dam with a bravado befitting the triumph of civilization:

> This morning I came, I saw and I was conquered, as everyone would be who sees for the first time this great feat of mankind. Ten years ago the place where we are gathered was an unpeopled, forbidding desert. In the bottom of a gloomy canyon, whose precipitous

> walls rose to a height of more than a thousand feet, flowed a turbulent, dangerous river. The mountains on either side of the canyon were difficult of access with neither road nor trail, and their rocks were protected by neither trees nor grass from the blazing heat of the sun. The site of Boulder City was a cactus-covered waste. The transformation wrought here in these years is a twentieth-century marvel.
>
> We are here to celebrate the completion of the greatest dam in the world, rising 726 feet above the bed-rock of the river and altering the geography of a whole region; we are here to see the creation of the largest artificial lake in the world – 115 miles long, holding enough water, for example, to cover the State of Connecticut to a depth of ten feet; and we are here to see nearing completion a power house which will contain the largest generators and turbines yet installed in this country, machinery that can continuously supply nearly two million horsepower of electric energy.
>
> All these dimensions are superlative. They represent and embody the accumulated engineering knowledge and experience of centuries ... The mighty waters of the Colorado were running unused to the sea. Today we translate them into a great national possession ... This is an engineering victory of the first order – another great achievement of American resourcefulness, American skill and determination.[51]

While there have been countless exceptions to hydraulic civilizations throughout history in the form of small urban communes existing across coastal areas and river valleys and self-governed as commons, they have received far less attention by anthropologists and historians who have been far more interested in unearthing the history of large-scale urban habitats attached to hydraulic infrastructures. The renewed interest in these alternative human communities was sparked in 2009 when the economist Elinor Ostrom became the first woman to receive the Nobel Prize in Economics for uncovering this rich history of governance, which often survived and flourished more successfully and for longer sustained periods of time by adapting human communities to nature instead of adapting nature to human communities. This renewed interest in commons governance is beginning to attract quite a bit of attention in scholarly circles and even in governance and civil society, as hydraulic civilization unwinds.

The melting of the Arctic and Antarctic, the emergence of powerful atmospheric rivers, changes in ocean currents, massive floods, extended droughts and heatwaves, spreading wildfires, and powerful hurricanes and typhoons are the manifestation of a rewilding hydrosphere that over the lifetime of today's children and well into the twenty-second century, will likely see the collapse of hydraulic infrastructure over much of the world and with it what we have come to know as civilization.

The question, then, is there some form of collective life that might emerge and perhaps even thrive alongside a rewilding hydrosphere? That prospect is already here in bits and pieces but yet to scale. We are on the cusp of a momentous transformation from a dying urban hydraulic civilization to an emerging "ephemeral society." Getting there is the big unknown.

Now, in the midst of a warming climate brought on by an industrial fossil fuel-based civilization, the fresh waters of the planet are fast diminishing all over the world. The rivers and lakes are drying up, and the dams and artificial reservoirs that make up a lattice-work of hydraulic infrastructure over every continent are passing. This reality becomes even more frightening when we stop to consider that 70% of all the remaining fresh water on the planet goes to irrigation. To get a handle on how dire the situation is, 20% of all remaining available fresh water on Earth resides in the five Great Lakes of North America: a chilling observation. In these end days of the hydraulic civilization, it's the major cereal crops – rice, wheat, maize, and soybeans – that consume 59% of all the available fresh water intended for international food crops.[52]

A recent segment on America's PBS public TV network reported on the catastrophic diminishment of fresh water on the planet with the stark headline that "as global groundwater disappears, rice, wheat, and other international crops may start to vanish."[53]

While the dark side of civilization has been widely scrutinized by post-modern scholars, particularly in relation to imperialist overreach and colonization, there is another less-explored aspect of civilization that has received very little, if any, attention at all. Scholars, of course, are quick to mention that dense urban life brings together otherwise unrelated individuals with diverse talents and skills, who service one another in a myriad of ways, providing the rudiments and gratuities of life. Anthropologists tell us that forager-hunter societies rarely exceeded one hundred or so blood-related and/or kinship-related extended families, making cooperation a collective endeavor. However, learning to live alongside thousands of unrelated individuals, of whom one knows little, requires new types of cooperation. To be "civil" is the hallmark of being civilized and it's about forging non-blood-related ties of sociability. In more advanced urban civilizations cooperation morphs into what's come to be called cosmopolitanism, learning to be tolerant of untold others. But it should be said that anyone living beyond the urban confine is likely to be thought of as savage and treated as alien others. Being civil, however, doesn't always come naturally and, for that reason, must be taught but also enforced by the establishment of laws, codes of conduct, and punishments, lest the fabric of civil society frays.

As alluded to earlier, there is, however, a deeper and more powerful tie that draws unrelated populations together as a social organism, which has received little, if any, attention among anthropologists and historians: that is, the extension of the "empathic impulse" in our neurocircuitry beyond our immediate extended families and neighborhoods to encompass diverse populations living side by side in increasingly dense and complex urban civilizations.

There have been to date three significant extensions of empathic consciousness, each embedded in the way our species has governed itself. Forager-hunter societies and later Neolithic farming communities confined their empathic impulse to blood relations and extended tribal communities. Unrelated human bands crossing into one's territory were often – but not always – considered alien "others" and a threat to their survival. Their psychic world is what anthropologists call "animist consciousness." They experienced a milieu populated by their ancestors in the netherworld, with whom they communed, and spirits of all kinds embedded in the mountains, rivers and streams, forests and savannas, with whom they related as beings.

The coming of the great hydraulic civilizations brought biologically unrelated people together for the first time. This tumultuous change shifted the empathic bond upwards to include "unrelated" people who came to be looked on as part of an "extended fictional family" attached to male gods from up on high in heaven, to whom they looked for guidance and came to obey as father figures and, occasionally, to a single male god overseeing the whole of his human family. This marked the beginning of "religious consciousness."

It's unlikely a coincidence that the axial religions – Judaism, Hinduism, Buddhism, Daoism, and later Christianity and Islam – appear around the same time and in the same places that hydraulic civilizations were beginning to bring together large numbers of unrelated people in dense urban environments. To take one example, in first-century Rome – the epicenter of a vast hydraulic civilization – tens of thousands of displaced migrant refugees from across the empire traversed Roman roads and found themselves alone in a capital city of more than a million people. Try to imagine their sense of isolation. No longer surrounded by blood relatives and extended kinship groups and separated from their spirit world, they forged a new parental bond with the prophet Jesus Christ, a father figure, whose unabiding love, nurture, and empathic embrace gave comfort, and came to see themselves as the children of Christ and related to other converts as brothers and sisters in the family of Christ. When Christian converts met each other on the streets of Rome, wearing identifiable garments, they would stop and kiss each other on each cheek and say hello brother and hello sister, acknowledging their common bond as Christ's children. And, like their animist ancestors, they would support one another and even give their lives for each other. This was the second great empathic shift.

The third great empathic extension, "ideological consciousness," emerged with the development of a fossil fuel-based industrial civilization overseen by nation states. The nation state became the new attachment figure. Unrelated local ethnic cultures, each with their own language and dialects and unique cultural traditions and ways of governance, were suddenly thrust into a vast new governing jurisdiction in the form of nation states. The primary mission of these newly conjured nation states, each presiding over more expansive hydraulic infrastructures, was to corral the disparate local and regional ethnicities across their governing domain into a single universal community. To accomplish this task, they invented a largely imaginary common heritage, along with a single shared language and a universal educational system, all designed to reorient these provincial groupings into a common fictional family under the wise and guided protection of their "motherland" and "fatherland." For example, when Italy was created as a nation state, the former premier of Sardinia joked that "We have made Italy. Now we must make Italians."[54]

This reshaping of human consciousness to form national families proved to be effective. Over the past two centuries, nation states emerged on every continent, each populated by fictional families, whose allegiance to the motherland and fatherland was reflected in their willingness to come to each other's aid and even fight and die "for their fellow countrymen."

Each of these new waves of consciousness – animist, religious, and ideological – served to expand the empathic embrace but often at the expense of viewing all other collectivities beyond their territorial boundaries as "alien others." Need we recall that countless numbers of our species have fought, subdued, enslaved, and killed their fellow human beings under the banner of tribal, religious, and ideological consciousness, and continue to do so to this very hour.

These empathic extensions in history come and go. Kinship groups, religious orders, and nation states rise and fall along with the communities attached to them. Today we are witnessing the early hours of a fourth stage of empathic consciousness – what a younger generation refers to as "biophilia consciousness" – in the form of an empathic identification with all that is alive on Earth. And many of the young are willing to dedicate their lives to caring for an ailing planet.

These epoch shifts in empathic attachment, at least up until now, have waxed and waned over history. There have been extended moments of empathic expansion, inevitably followed by carnage and collapse of entire civilizations. There are at least two primary reasons for their mercurial coming and going. First, every civilization up until now has had a life span – a birth, growth, maturation, and death. Their demise is often the result of military conquest, a dramatic change in the climate, or the inevitable entropic bill. Secondly, and less obvious, each new empathic step forward toward a more universal intimacy is inevitably challenged by previous empathic attachments. Today, as a younger generation lurches toward biophilia consciousness and a planetary embrace it's perceived as a threat to older forms of empathic attachment who feel their worldview is being lost in the shuffle, leaving them isolated and abandoned. And they fight back to hold onto their allegiances for fear of losing their identities.

Right now, tribal blood wars are raging, religious wars are occurring, and ideological wars are proliferating, and everywhere people are dying, communities are being upended, and ecosystems are faltering. This sad state of human affairs is occurring even as the realization sets in, at least among the young, that a warming climate brought on by climate change is taking the whole of the human population and our fellow creatures to a mass graveyard.

The irony of it all is that each phase of empathic attachment over history is, in its own way, a search for universal intimacy – an often-unrecognized quest to be at one with existence itself. A skeptic might argue, with some justification, that the very idea of universal intimacy is itself an oxymoron. How can intimacy and universality travel together? Perhaps so, but it's the gnawing underbelly of consciousness, always lurking in the background.

Johann Wolfgang von Goethe, the great German philosopher and scientist of the Romantic Period in Germany, put it best. Of nature he wrote, "we are surrounded and embraced by her – unable to step out of her, unable to penetrate her more deeply."[55] Goethe came to believe that while every creature is unique, each is connected in a single unity. Goethe argued that "each one of her creations has its own character . . . that altogether make one."[56]

Goethe wrote of the universal drive of empathic embrace even before humanity had devised a word to describe it. He explained, "to find myself into the condition of others, to sense the specific mode of any human existence, and to partake of it with pleasure" is to affirm the universality of life.[57] He summed up his sense of universal intimacy, suggesting that it is the "beautiful feeling that only mankind together is the true man and that the single individual can be joyful and happy only when it has the courage to feel itself as part of one."[58]

CHAPTER 3

Gender Wars

The Struggle Between *Terra Firma* and Planet Aqua

The hydraulic civilizations took our ancestors away from a highly localized and distributed way of living off the land during the whole of the Neolithic era. This earlier era, with its small-scale agriculture and pastoralization was more or less egalitarian, with women attending the fields and the crops and men hunting prey and overseeing the herds. If there was any distinction to be made, it was that women played the dominant role in the subsistence economy, as diets were more plant-based than animal-dependent. Then, too, women were largely responsible for originating economic life by their invention of preservatives to store meat, pottery to store grain, and the tanning of hides to provide garments and wrappings for year-round shelters.[1] Life expectancy of our ancestors who lived in the Neolithic era was between 25 and 28 years of age.[2]

Picking Sides: The War of Goddesses and Gods

Neolithic cultures everywhere paid homage to goddesses often in the form of female serpents, whose gestational waters fertilized the fields and provided the nutrients to sustain life. The small-scale Neolithic agricultural economy was one of constant care and nurturing and socially constructed as a commons to ensure that the community and its heirs would not deplete the generative forces that provisioned its human domesticators.

While in previous times, the waters were often associated with female attributes of fluidity, fertility, and generativity and ruled by female deities in the form of serpents attached to rivers, lakes, and streams, the advent of highly centralized and bureaucratic hydraulic kingdoms ruling over populations identified more with the male attributes of power, subjugation, protection and punishment, and even control of temporality and spatiality (the first calendars and laws governing the propertization of nature emerged at this time in history). Having tamed the waters, allegiance to the earlier water goddesses steadily gave way to faith in mostly male gods overseeing their earthly creation from above. The rise of hydraulic civilization tipped the scale to male-dominated deities and has remained so into the modern era. The ongoing battle between civilization and nature came in the form of a life-and-death struggle between male and female agency, which is still in place millennia later.[3]

The earliest recorded histories of hydraulic civilizations tell the story of male super gods slaying female serpent water goddesses. In Babylonia, Marduk slayed the female serpent, Tiamat.[4] In Greek mythology, Zeus defeated Typhon, the serpent child of the goddess Gaia, securing the reign of the Olympian gods.[5]

While the waters were perceived as the womb of life that animated all of existence among our Neolithic ancestors, the coming of hydraulic civilization and urban life tipped the scale to a male symbology and cast the waters as a menacing presence that needed to be tamed and domesticated. We need to be reminded that for the most part the male heroes in mythology and across antiquity – Beowulf, Perseus, and Jason and the Argonauts – were hunters seeking the annihilation of the deep and the feminine persona. William Young and Louis DeCosta's study of "water imagery in dreams and fantasies" in the journal *Dynamic Psychotherapy* gives a chilling account of the male heroes of mythology hurtling into the depths to stalk and kill the goddesses of the water. They write:

> When they draw upon the heroic plunge into the watery abyss to contend with sea monsters, as in the story of Jonah and the Whale or Beowulf, are they not also evoking the heroic venture into the 'watery' abyss of the unconscious? Or, similarly, is not the heroic mastery of forbidding waters with their effeminate allure a reference to the male's unconscious fear of a drowning engulfment by a

> woman? . . . The feminine and water appear to be virtually merged as representations . . . as seductive and destructive mermaids and sirens; the ocean as life-giving mother; the Arthurian lady of the lake.[6]

Endless origin stories tell a similar tale of an earlier animist culture with female water deities, often in the form of serpents, being put to their deaths by powerful male gods overseeing hydraulic civilization.

Veronica Strang, a professor of anthropology at Durham University, makes the point that water cults generally overseen by female goddesses existed across Europe as far back as 6000 BCE. Although the rise of hydraulic civilization and male control of the waters in the Mediterranean and Europe tipped the scale to male gods overseeing the deep, it was not without a protracted struggle.

Water worship of female goddesses at the site of holy wells was commonplace. Springs and particularly wells were viewed as "the exit from the womb of the Earth Goddess," who they looked to as the incubator of life.[7] Annual pilgrimages to these holy sites were an integral part of the societal life and times and an embedded centerpiece of how our ancient brethren viewed existence. Millions of our forebears sought relief in visits where they paid homage and engaged in intimate rituals calling on the goddess of the waters to share her generosity by granting her worshipers fertility, regenerativity, healing, and good health. As late as the Christian era there remained thousands of holy wells across Europe, and routine pilgrimages were undertaken to these sacred sites to experience their "healing properties," posing an embarrassment to the Church. In the sixth century, Pope Gregory had enough and decreed that instead of destroying the holy wells, they be converted to Christian shrines overseen by phallic male symbology. In Britain and Ireland alone, "thousands of holy wells . . . were thus subjected to saintly takeovers and makeovers."[8]

The struggle between a male-dominated Church presiding over God's creation from above and these earlier religious cults and rituals paying homage to female goddesses of the deep continued, with Church decrees issued time and again forbidding worship at the sites of holy wells as late as the twelfth century and even beyond. The long-held practice of worshiping at wells is suggestive of the great undercurrent that has marked the human journey since the dawn of hydraulic civilization 6,000 years ago, pitting *terra firma* over fluid waters and male dominance of a passive nature over female generativity.

Strang tells the story of the Saxons' pagan rituals honoring "the spiritual power of water . . . [recounting that] they used water from sacred springs in charms and ritual practices and believed that drawing water at sunrise and in silence from an eastward flowing stream" would "restore growth, heal eruptions, and make cattle strong." Eventually, some give-and-take was granted by the Church by at least renaming the holy wells for female saints. But most Christians still regarded the wells "as healing wells" that could "cure ills and assist fertility far into the seventeenth century" and the dawn of the Enlightenment with its newly rationalized worldview of Baconian science and universal mathematics riding alongside a new era of modern hygiene and public health.[9]

Still, even today, the feminine gender footprint is discernable at every turn. The River Dee in Britain is named after Deva, the Old English word for goddess. The Seine River in France is named after the water goddess Sequana, a Celtic deity who inhabited the waters. The triumph of male gods over female goddesses was intertwined with the control and propertization of the waters at the inception of the great hydraulic civilizations in both the West and the Oriental world, with men overseeing the cultural norms, the rules of social engagement, the economic parameters, technological deployments, and governance, and women left to the nurturing of offspring and caretaking of domestic life.

It's interesting how little attention has been given to how our species' relationship to the waters has underwritten gender relations that establish feminine and masculine stereotypes, portraying the feminine as passive, receptive, emotional, nurturing, and sacrificing, and the masculine as active, aggressive, powerful, rational, and detached. These stereotypes largely determined the role that males and females would come to play in the evolution of society. From the early centuries of hydraulic civilization, it's the male cohort that has conceptualized, deployed, and managed the massive hydraulic infrastructures from the first manifestation of domesticating the waters in ancient Mesopotamia to the near-total privatization and commodification of the waters in the industrial hydraulic era in the nineteenth and twentieth centuries.

If the masculine fear of the deep has been a recurring theme that has complicated our species' relationship to the waters, it has been met with an equally strong love affair . . . but with conditions. At the height of the Roman Empire, for example, the waters were a defining element of its rise to power and majesty. Although we all know of the oft-heard phrase that "all roads lead to Rome," it would be more accurate to say that "all aqueducts flowed to Rome." The Roman Empire surpassed all earlier civilizations in its harnessing of the waters. From a very practical point of view, housing upwards of one million inhabitants – free, indentured, and enslaved – also required bringing a massive volume of water to Rome from across the hinterlands as well as transporting water to all its governing regions throughout the empire that stretched from Greece to Western Europe, the Balkans, the Middle East, and North Africa. While the upfront story heaps praise on the elaborate labyrinth of road systems that connected the empire, it was the great aqueducts that rechanneled the rivers of three continents across an immense kingdom that was equally important in uniting its millions of citizens.

Hydraulic engineering was the pride of the empire – an enormous feat of human ingenuity and technical prowess designed to literally tame the waters across a vast landscape. But Rome's attachment to the waters is even more complicated. Its engineering prowess was a double-edged sword. While capturing and enslaving the waters was carried out with the same military precision it employed in capturing lands and peoples, its desire to be at one with the waters was an equally compelling counterforce. And here's where the contradiction comes in. If the ubiquitous presence of Roman fountains spouting across the empire represented the phallic libido of a male-dominated civilization, luxuriating in the tens of thousands of elaborate baths and pools, likely reflected an unconscious desire to be swallowed back into the watery bonds of earlier pre-memories in the womb.

Majestic baths and pools were ever-present across the Roman Empire, attached to military barracks, visible in public squares, and embedded in private villas. The numbers tell the story. Consider that the city of London in 2008 had only 196 public swimming pools to serve a population of 7.7 million people, while Rome, at its height, boasted 800 public pools open to all its citizens in its heyday, with a population of around one million.[10]

The British historian, Edward Gibbon, recounts that the "meanest Roman could purchase with a small copper coin the daily enjoyment of a scene, of a pomp and luxury that excited the envy of the kings of Asia."[11] Meanwhile, the emperors, aristocracy, and wealthy merchant class of the time sported majestic pools and baths at their estates, but not without collateral damage. The Emperor Diocletian's bath and pool was erected over seven years by indentured Christians who were summarily put to death upon its completion.[12] Swimming throughout the Roman Empire was a universal pastime, and even an obsession – so much so that when the word got out that the Emperor Caligula did not know how to swim, he became the subject of public ridicule.[13]

Rome's governing elite were of a single mind in the conviction that sequestering and controlling the waters for exclusive human consumption, irrigation, and sheer pleasure was their greatest accomplishment. When Sextus Julius Frontinus, lead architect and engineer of the empire, surveyed the 441 years from the founding of the city to the present with the final completion of nine aqueducts, which brought the waters from distant parts of the Alban Hills to the people of Rome, he quipped, "with such an array of indispensable structures carrying so many waters, compare, if you will, the idle pyramids or the useless though famous works of the Greeks."[14]

To be clear, the Roman Empire was a polytheistic culture and worshipped both male gods and female goddesses. Jupiter, however, the sky god, was regarded as the supreme "overseer [of] all aspects of life... including protecting the Roman state." Female goddesses, not surprisingly, played more of a subservient role. Jupiter's wife, the goddess Hera, kept a "watchful eye over women and all aspects of their lives."[15]

This story of the controlling lover seeking both mastery and intimacy with the planetary waters, though ultimately at odds, has been a defining dialectic in the history of our species' relationship to the Earth's hydrosphere since the birth of civilization. The historian Brenda Longfellow, of the School of Art and Art History at the University of Iowa, reports that over the first century BCE, Rome's powerful political elite, military leaders, and wealthy merchants began showcasing their opulence by creating magnificent villas surrounded by artificial landscapes that mimicked nature's own – each more opulent than the other – a kind of masterly game or sport, the aim being to outdo one another in remaking nature, each in their own male image; just what one might expect of a man-God. Longfellow reports that the rich and powerful in Rome "took great joy in shaping the topography of the landscape, creating rivers, waterfalls, and hillsides where there were none before" – with a keen focus on water features "that ranged from evocative canals to artificial waterfalls fed by aqueducts or channels directed into natural or man-made caves."[16]

The centerpieces of these remakes of nature were elaborate water displays, each more splendid than the other, featuring fishponds and water alcoves and, at the apex, spectacular fountains. The not-so-subtle inference was that the elite of Rome were male gods, each a master of the waters and the creation. Today's feminist historians would add that what was taken for granted by the male gods' mastery of the waters was the banishment of the once-powerful female goddesses. They had no place in this new male-made water paradise. Of course, the sea change from the female goddesses of the watery womb to the male/man gods with their phallic fountains and male-made nature was likely subconscious, but no less a powerful turn in the way subsequent generations would come to perceive nature.

Confident of their new status, Rome's elite began to share their paradise with the public at-large. Longfellow recounts the rivalry of Julius Caesar and Pompey, Rome's most powerful rulers, who began to compete with one another by developing elaborate public gardens featuring water parks every bit as grand as their own private gardens. Longfellow noted that "Pompey offered spectacles to the average Roman that they had heard of but perhaps had never experienced firsthand – and now, thanks to Pompey, they had access to the amenities and life of leisure associated with villa culture." Not to be outdone, Caesar built his own public complex – the Forum of Caesar – which includes a monumental fountain, the Appiades Fountain, whose centerpiece was a "sculptured display of the Appian Nymphs" – the mythical female spirits of nature that were thought to inhabit the rivers, lakes, and streams, but were now showcased as wards of a male-dominated fountain of life.

Both rulers were mindful of the benefits of sharing their public gardens of paradise with the Roman public as "a staging ground for paying off votes" and "fostering goodwill." But it was the Emperor Augustus who fully understood the implications of providing elaborate over-the-top water displays. Longfellow sums up the political and sociological significance of these public gestures:

> Augustus recognized the public appeal of green spaces and associated artistic water displays and, as part of his efforts to create a new political stability in Rome, he sanctioned civic green spaces dotted with water spectacles for the leisure activities for the general public. Moreover, it was under Augustus that the monumental decorative civic fountain was transformed from being one amenity among many set within a large space to a monumental stand-alone edifice that could telegraph the ideology of the emperor.[17]

The decline of the Roman Empire beginning with the split into Eastern and Western empires in 286 CE, followed by a blitz of Gothic invasions and, as mentioned earlier, a sack by the Vandals in 455 CE, marked the beginning of the end of the signature hydraulic civilization up to that time in the Western world.

While the Roman Empire collapsed in 476 CE, its draconian male-dominated worldview has lived on in myth and practice, flaring up in brief intervals with often horrendous consequences. It was not surprising, then, that in the midst of a global warming crisis threatening the extinction of life on Earth, and the related gender wars over male dominance and aggression manifesting itself across the social spectrum, from a militarized geopolitics on high to the "#MeToo movement" below, revealing deep-rooted gender inequalities in the economy and even family affairs, that the Roman moment has burst onto the global public stage once again, opening up a can of worms.

It all unfolded when a Swedish influencer Saskia Cort asked her Instagram followers, "to ask their male partners and friends how often they thought about the Roman Empire."[18] Women were surprised when men everywhere replied that they think of the glory days of Rome "three to four times per month," "every couple of days," and "at least once per day." The responses went viral on TikTok, exposing a new moment in the gender wars, with men glorifying the military prowess, engineering achievements, and bureaucratic acumen of the male-dominated empire, not to mention its generosity in providing "bread and circuses." The resurrection of the Roman moment spilled out onto the lead pages of *The New York Times*, *The Washington Post*, and other major media, still another example of the staying power of patriarchy that has traveled alongside hydraulic civilization for six millennia of history.

The ascension of the Catholic Church and the Holy Roman Empire in ensuing centuries after the Fall of Rome ushered in a more loosely hung decentralized governance, with modest control by the Church hierarchy. The great Roman aqueducts, like the Roman road systems that crisscrossed Europe and extended into Africa and Asia, were left unattended and, for the most part, abandoned; and with that the fetishization of the waters fell by the wayside, as the Church turned inward and landward, focusing attention on *terra firma*. The waters were seen as the void, as in the earlier Hebraic cosmology, and to be avoided rather than idolized. Water nymphs were erased from the Catechism as all eyes were focused on the everlasting life that awaited the faithful in the next world.

Little has changed. With the minor exception of the Gnostic Christian cults that flourished for a short time between the crucifixion of Christ and the rise of the Holy Roman Empire when the female persona was elevated in status along with a very different idea about the meaning of the resurrection and the idea of Christianity, the female gender continued to be marginalized and constricted.

The Roman Catholic Church then and now, 2,000 years later, only sanctions males as priests, cardinals, and popes to serve as God's messengers to the faithful, and reminds both genders that God created Adam first and extracted a bone from his rib to fashion Eve. And, as to women's degraded status, the gender was disenfranchised from the start, with the biblical account of Eve disobeying the Lord's order not to eat the apple from the tree of knowledge of good and evil. Her indiscretion was met with the Lord expelling both Adam and Eve from Paradise and sentencing all of their heirs thereafter to a life of toil and hardship.

Often overlooked is that the decline, disrepair, and collapse of the great hydraulic infrastructure that was the keystone foundation of Roman culture, governance, and economic life also led to the partial abandonment of water imagery as a defining feature of gender relations in subsequent centuries. Aqua metaphors became less appealing than *terra firma* metaphors in describing gender relations. While the female gender was still diminished and looked upon as passive property, the waters dried up as a carrier of gender categories.

That all changed again in the sixteenth century with the onset of the great oceanic explorations, recasting gender relations once more. The race to explore, subdue, and colonize entire "new worlds" and whole continents hoisted the male figure as the master of the seas and cast the female figure as the seductor of the seas, lying in wait in the deep to swallow the invaders.

Gender relations were further ossified during the Romantic Period, which stretched from the 1790s to the 1850s, becoming the political playground for a repeat of the historical struggle around gender roles and the waters. A male cast of poets and authors – mostly English and Scottish – began seesawing back and forth between the desire to vanquish the feminine persona, while holding on to a primal attachment to the womb – a resurfacing of the love/devouring dynamic that has long gone unspoken but nonetheless been a recurring theme and presence in the male/female gender dynamic during the whole of the history of hydraulic civilization over 6,000 years.

Lord Byron was among the towering figures of the Romantic Period and today might be described as a macho Romantic. And his "love affair" was with the waters. A contemporary of Percy Shelley and a personal friend, Byron showed his disdain for what he viewed as the feminized Romantic poets of his day, including Shelley. Byron's first love was the waters. Well known at the time, he was revered across Europe for his swimming prowess. He often chided Shelley for his swimming skills – in fact, Shelley eventually drowned when a boat he was on capsized in the Gulf of La Spezia.

Lame and disfigured in childhood, for Byron swimming in the waters became his refuge and freedom. He remarked, "I delight in the sea and come out with a buoyancy of spirits I never feel on any other occasion," and "if I believed in the transmigration of souls, I should think that I had been a 'merman' in some former state of existence."[19] Byron's disparagement of the feminine gender was demonstrable. When on a trip to Ithaca and asked by his hosts if he'd like to visit their local antiquities, he let go his disdain for the feminine mystique, muttering, "do I look like one of those emasculated fogies?" upon which he added "let's have a swim."[20]

There is no reference in Byron's poems and satire about what lay below the surface of the sea, and he never referred to the world of the deep. Rather, his sole interest was conquering the waters. Other poets and literary giants of the Romantic Period, including Samuel Taylor Coleridge and Robert Browning, were of like mind. Charles Sprawson, the author of *Haunts of the Black Masseur: The Swimmer as Hero*, observed in his study of avid swimmers throughout history, but particularly during the Romantic Period, that they were like Narcissus, fixated with their own self-importance, entitled, preoccupied with power, arrogant, and showed a lack of empathy.

While the Romantics took aim at the rational detachment of Enlightenment thinkers with their loathing for nature and the physicality of existence, they were, for the most part, of a mind that nature was to be reveled in, but always with a caveat that it was to be regarded as a gift to delight. As to how the female gender fit in, given that the female and nature were long looked on as "the other," most of the dominant male figures of the Romantic Period – in both poetic verse and the arts – continued to place the feminine persona in an inferior rank, whose primary purpose was to procreate, domesticate, and serve a male mystique.

For the Romantics, freedom meant autonomy, and the female represented "the devouring other." Anne K. Mellor, a professor of women's studies at the University of California, Los Angeles, captures the dynamic as seen from the eyes of male poets and artists of the Romantic Period. Focusing on William Blake, one of the earliest Romantic writers, she notes:

> Blake's gender politics conform to those of the other Romantic poets: the male imagination can productively absorb the female body, but if the reverse occurs, as when Vala [a prophetess imbued with mystical energy] or the Female Will covers the body of Albion [a poetic name for England] in her veil, the image is negatively equated with a fall into death and self-annihilation.[21]

Blake was not alone. The poet, John Milton, the Romantic Period's guiding light, was quite clear on the question of the relationship between male and female. He wrote, "who can be ignorant that woman was created for man, and not man for woman?"[22] What Milton failed to share is that his quote is reminiscent of an earlier quote by Francis Bacon, the father of modern science, who remarked that "the world is made for man, not man for the world."[23]

At least Milton and several male Romantics were willing to soften their stance on the role of the female. In his prose, Milton makes clear that women should not be regarded as chattel but as "a spiritual helpmeet" whose primary mission in life is to enhance his own autonomy and mastery. Milton wrote, "man is not to hold her as a servant but receives her into a part of that empire which God proclaims him to, though not equally, yet largely as his own image and glory."[24]

It should be noted that the macho approach to the vast oceans during the Romantic Period was countered by a more genteel attachment. The lake poets were more provincial and even made room for the female goddesses – the nymphs of the rivers, streams, lakes, and wells. However, the more expansive philosophers, poets, artists, and essayists of the period played in a bigger arena – the great oceans – and were wedded to a global reach in which the island nation was the command center for what would become the Age of Imperialism, and this arena was viewed as the "preserve of a male elite."[25]

For the record then, the Romantic poets, writers, and artists of the day reveled in a masculine handling and taming of what Wordsworth in "The Blind Highland Boy" refers to as "the wonders of the deep."[26] Conquering the deep became the overarching theme of British imperialism in the nation's quest to colonize the world. And the Romantics brought with them a soft arsenal of poetry, literature, and the arts to accompany their hardcore weapons of war all in the form of a romantic adventure aimed at sequestering the deep and bringing civilization to the world. In James Joyce's twentieth-century novel, *Ulysses*, a Mr. Deasy remarked, "the sun never sets on the British Empire." The German Romantics, by contrast, whose coastlines never abutted the Atlantic Ocean, were less able to imagine an oceanic sublime over which they could preside. While Germany, too, dreamed of colonizing the world, the nation was only able to extend its global reach to a handful of colonies in Africa and the Pacific Ocean.

Lord Byron captured the spirit of the times and the essence of British Romanticism and its imperialist ambitions in his poem The Corsair . . .:

> O'er the glad waters of the dark blue sea,
> Our thoughts as boundless, and our souls as free,
> Far as the breeze can bear, the billows foam,
> Survey our empire and behold our home!
> These are our realms, no limits to their sway –
> Our flag the sceptre all who meet obey.
> Ours the wild life in tumult still to range
> From toil to rest, and joy in every change.
> Oh, who can tell? not thou, luxurious slave!
> Whose soul would sicken o'er the heaving wave;
> Not thou, vain lord of wantonness and ease!
> Whom slumber soothes not – pleasure cannot please –
> Oh, who can tell, save he whose heart hath tried,
> And danc'd in triumph o'er the waters wide,
> The exulting sense – the pulse's maddening play,
> That thrills the wanderer of that trackless way?
> That for itself can woo the approaching fight,
> And turn what some deem danger to delight;
> That seeks what cravens shun with more than zeal,
> And where the feebler faint – can only feel –
> Feel – to the rising bosom's inmost core,
> Its hope awaken and its spirit soar?

Although a handful of other Romantics – mostly women – attempted to resurrect the waters of the deep and the image of the feminine as the primal life force, they were brushed aside in the second half of the nineteenth century by a manic hysteria over the newly discovered dangers to public health from polluted waters. The discovery that typhoid fever and cholera – both deadly diseases – are transmitted to humans via polluted water created near hysteria in Britain, France, and across Europe.

Béatrice Laurent, a professor of Victorian studies at the University of Bordeaux-Montagne in France makes the point that the fight between the feminized waters and a masculine *terra firma* suddenly metamorphized into a life-and-death struggle. Harking back to the belief that menstruation and the monthly shedding of blood – the so-called curse of Eve – made the female the impure carrier of disease, it wasn't long before popular lore began to connect the female with polluted waters. She wrote, "because women and water were considered equally susceptible to pollution, and because of the supposed permeability of the female body and character, it was difficult to determine which of the two caused the adulteration of the other." In short, the female mirrored "brute nature."[27]

And the "findings" of modern science and medicine didn't stop with just the menstruation scapegoat. The ancient "Hippocratic theory of humors" was resurrected in a new guise with some voices in the medical community "suggesting that improperly regulated body fluids provoked the most feminine of diseases, hysteria." The new field of hydrotherapy stemmed from the belief that fluids could be regulated with water. A noted physician of the day suggested that "the douche is a very necessary part of the treatment."[28]

In this instance, "ideas have consequences" even if ill-informed. Elaine Showalter, a literary critic and professor emeritus of Princeton University, points out that this updated "medical" belief of the female persona and impure waters metamorphized and became a critical underpinning of psychiatric theory and practice. She reports that "at the end of the nineteenth century, hysteria, the classic female malady, became the focal point of the second psychiatric revolution."[29]

While the popular diagnosis tying the female body to polluted waters and the carrier of deadly diseases was ludicrous, but nonetheless widely accepted, the discovery that the waters were the carriers of life-threatening diseases propelled the modern movement of sanitation, first in Britain, and shortly thereafter in France, Germany, the United

States, and the rest of the world. In the second half of the nineteenth century, the purification of waters became the centerpiece of a great infrastructure revolution, with the introduction of water purification systems and the piping of purified water directly into homes and businesses, accompanied by the build-out of sewer systems.

The United Kingdom led the way with the passage of the Metropolitan Water Act of 1852 establishing strict standards to govern water quality. Other nations followed, provisioning pure water for consumption. Still, 171 years later in 2023, two billion human beings lack access to safely managed drinking water, and 4.2 billion people – half the world's population – go without sanitation in their homes, and three billion people lack even basic hand-washing services and do not even have access to soap.[30] The result is hundreds of millions of human beings suffer from waterborne diseases each year.

The capture, pacification, purification, and distribution of fresh water and recycling of wastewater in the late nineteenth century and early twentieth century in Britain, and from there the rest of the world, marked the highpoint and endgame in sequestering and domesticating our most valuable lifeforce. The sequestration of fresh water, in turn, became the integral component of the industrial revolution and the water–energy–food nexus became the workhorse of the capitalist system in both market-driven and socialist countries.

But now, the waters have had enough. The planetary hydrosphere is reeling and no amount of human intervention to shackle it is any longer possible. The question, in all of its various dimensions, is whether we will relearn how to be at one with the waters as a steward and adapt to wherever it takes us, rather than the other way around. Much will depend on how we go about recasting gender relations in the Anthropocene.

Women: The Carriers of the Water

All through history and across cultures women have been the carriers of the water. Even today, millions of women spend several hours a week walking miles on end to retrieve water and carry it home. In some regions in Africa and Asia, women walk on average 3.7 miles per day for water. The heavy load, generally balanced on top of their head, often leads to serious injuries. A recent study by the United Nations International Children's Emergency Fund (UNICEF) and the World Health Organization (WHO) reports that 2.1 billion people – mostly women – continue to be the carriers of water.[31] Although little discussed, these long daily treks to retrieve water have been a decisive factor in preventing young girls in developing countries from attending school and receiving an education, sealing their fate by keeping them domiciled to the hearth and home for the remainder of their lifetime.

When my mother was born in 1911, no woman in any single sovereign nation of the world had the right to vote. Norway became the first independent nation to grant women the vote in 1913. The United States government granted women the right to vote in 1920.[32] The other nations of the world followed throughout the remainder of the twentieth century. While women have belatedly won the right to vote, they have been systematically denied inclusion in decision making till late, particularly in the corridors of governance and the executive boardrooms of the corporate economy – less so in civil society. And nowhere is the divide between male dominance and female subservience more pronounced than in the arena of global water politics and policies – especially in the poor countries, but also some of the wealthiest nations in the world.

Since the hydrological cycle is the agency that generates life on the planet and assures a flourishing milieu, the near exclusion of the female gender from decision making affecting the waters speaks volumes to the male stranglehold over the hydraulic infrastructure that controls all of life – a position of dominance that until the past several decades has gone undiscussed and swept under the proverbial carpet by governments everywhere and by global governing institutions and the capitalist economy writ large.

What's indisputable is that the planet's hydrological cycle and the whole of the hydraulic infrastructure is controlled almost exclusively by men as a male domain, as it has been since the rise of the first hydraulic civilizations several thousand years before Christ.

Margreet Zwarteveen, an irrigation engineer and social scientist and professor of water governance, is one of the very few women with firsthand experience of working within the top echelons of the hydraulic field and provides an insider's look at this male-dominated sector. She points out that the hydraulic profession, which controls the whole of the waters used by our species, is roughly divided into two levels: the water professionals and the international group of experts who oversee the knowledge of irrigation systems and whose expertise is embedded in world governing bodies, including the UN, World Bank, IMF, OECD, and related think tanks, research organizations, consulting bodies, and universities; and on a parallel and adjoining track, the millions of individuals who plan, deploy, and manage all of the components of the global hydraulic infrastructure:

> Both groups are male dominated in that there are many more men than women employed as irrigation researchers, experts, engineers, planners, or managers. Indeed, the professional involvement with irrigation, be it as an engineer, manager, planner, or researcher, is (or at least used to be) very much identified and perceived as a male activity.[33]

Although not openly aired, few would dispute the claim that of all the components that weld our collective humanity together locally and globally, the hydraulic infrastructure is the underlying girder upon which the entire technological civilization rests. The oft-ignored truth is that this system, which has been with our species since the rise of the first hydraulic empires, is one of control and domination. And what of the technical competence that goes in tandem with hydraulic civilization?

A growing number of feminist scholars would argue that this engineering prowess "is central to the dominant cultural ideal of masculinity and its absence a key feature of stereotyped femininity."[34] Sequestration, domination, and control of the waters would be impossible without companion military bureaucracies to pacify populations, transform the geographic domain, and protect the infrastructure from seizure and destruction by hostile forces. Of note, all of the colonial regimes of the nineteenth and twentieth-century industrial era – the British, French, Spanish, Portuguese, Dutch, German, Japanese, and U.S. – which seized territories across all of the continents, came with two sets of boots on the ground: the military command and hydraulic engineers. They each brought with them the rationalized bureaucracies and command and control traditions of a male-oriented culture that's been an immovable presence since the first hydraulic civilization emerged in ancient Mesopotamia.

Their presence and penetration across the continents, while not welcomed by the indigenous populations, who rightfully saw them as invaders, were nevertheless heralded in their homeland as the carriers of enlightened civilization to the backward peoples of the world. The Dutch colonial engineers, the masters of hydraulic infrastructure, were idealized back home as the great unsung heroes of civilization itself:

> engineers were involved in a heroic struggle to conquer challenging tropical water streams. They did this far away from home, in a deadly climate, and in circumstances that were also extremely harsh otherwise. Especially the pioneers among the builders of monuments were true heroes in the eyes of later generations.[35]

Similar stories of heroic male hydraulic engineers who brought civilization to the backwaters of the Earth are recounted by every great power invested in hydraulic colonization – all in the name of modernization and advancing the Age of Progress. These paeans to the hydraulic engineers and the masculine mystique that accompanied the profession attracted a generation of young boys in the twentieth century, eager to be among the foot soldiers of a hydraulic future, bringing enlightenment to the masses. The indoctrination into the hydrological profession served to further the masculine persona – while associating "the other" with the female and the feminine, the passive and the dependent, the emotional and dreamy, locking generations of women into an inferior status as uneducable and reduced to the role of custodians of offspring and carriers of the water.

The tragic toll on the human psyche of this longest sustained psychological genocide in history – the mass extermination of the female persona in the public realm – robbed our species of the nurturing, empathic, and generative qualities of life, leaving our collective humanity prey to a cold, calculating, rationalizing, and detached approach to interacting with the planet's hydrosphere.

Nowhere are the masculine persona and feminine dependency more widely adhered to, although cloaked and subtle, than in the board rooms and executive suites of the world's leading governing institutions – the World Bank, OECD, IMF, et al. Now, however, male hegemony over the waters is beginning to be challenged. Over the past four decades or so, a new generation of women water activists, mostly in the developing nations, have begun to speak up and are demanding an equal role in determining our species' relationship to the waters, especially now that we find ourselves caught up in a global hydrological cycle that is rewilding, with consequences that will impact the very viability of life on the planet – and not in the distant future, but in the coming decades.

The opening salvo in women's struggle to claim a seat at the table and a voice in the administering of the waters couldn't be starker in contrast to what has been a masculine milieu for millennia. Activist women across the world "argue that debates about water management highlight the importance of building institutions that move beyond 'efficiency' to embrace and engender an ethic of care."[36] While women water activists acknowledge that global governing institutions like the World Bank, UN, IMF, and OECD are beginning to talk the talk and even include in their official reports and policy statements a need to involve women in the decision-making process regarding water policies, in practice they continue to promote a privatization of the waters – what's called public–private partnerships – handing over day-to-day control of administering the waters to an elite group of global corporations, often in the form of subsidized long-term leases, giving them carte blanche to oversee water deployment. These so-called public–private partnerships allow the male persona to continue to dominate the field, assuring that cost-benefit analysis, quarterly profit-and-loss statements, and return of revenue to shareholders is the line never to be compromised, while suggesting that the male-driven approach is the best way to assure the equitable sharing of the waters.

Feminist water activists would counter-argue that regarding the waters, "caring practices give primacy to relationships among all those human and non-human entities with whom one is in relationship."[37] The gender divide around our species' future relationship to the waters couldn't be clearer. The male perspective views water as a "resource," while the female perspective considers the waters to be a "life force." The male approach is to corporatize and commercialize the waters, while the female approach is to regard the waters as a shared global commons. These two very different ideas on how to interact with the Earth's hydrological cycle define the great struggle ahead. The outcome will likely determine the very future of life on the planet.

Does anyone really believe that human engineering can hold the waters back: that we can continue to sequester them, tame them, modify them, and manipulate them at the bidding of governments and the whims of the global marketplace? How likely is it that the male-dominated hydraulic civilization, as we've come to know it over six millennia of history, will remain viable with an unbridled hydrological cycle that's changing the topography, geology, and physiology of the Earth's ecosystems and reshaping the evolution of the planet's three other spheres – the lithosphere, atmosphere, and biosphere?

Learning to live in a fast-evolving hydrosphere will require a vigilant and empathic sensitivity to the freeing of the waters, carefully listening to what the waters are telling us. That listening calls on other qualities of our species' innate biology. Our ability to cooperate with our fellow creatures and share in a global water commons will offer up the new opportunities for life in the future. These are the qualities that we traditionally associate with the feminine persona, long held hostage and locked away over the span of hydraulic civilization. But we know that the empathic impulse is wired into the neurocircuitry of every male and female born of our species. If the Age of Progress represented the high point of a distorted concept of masculine identity bound up in controlling the hydrological cycle of the planet, the emerging Age of Resilience positions the feminine persona of empathic engagement as a counterweight and a biological phenomenon shared by both male and female babies. Ensuring that the empathic impulse wired into the neurocircuitry of every baby is nurtured will require a more enlightened social narrative that allows this most basic of all human impulses to mature and flourish in the form of biophilia consciousness as we adapt to a rewilding hydrosphere that's remaking the natural world.

Much will depend on our willingness to surrender to the idea of unknowability when it comes to the complex ways our hydrosphere evolves, changing the dynamics of life on Earth in a constant renewal. That doesn't mean de-learning about how the hydrosphere functions and taking note of the ways it has developed over the past two billion years. On the contrary, what it does mean is that the hydrosphere is a self-organizing and evolving sphere continually animating the entirety of the Earth in ways that are both subtle and expansive and impossible to predict with certainty in advance or be reined in and subdued.

At best then, we can learn from the past and anticipate where we might be heading but, more importantly, develop a phenomenology of adaptivity that will allow us to continually "stay afloat" wherever the hydrosphere takes us. To "be of the waters" is to steady our course and secure our place in a complex living organism, of which we are but one of the players, whose own resilience and well-being depend on "going with the flow" with a sense of critical thinking skills and technological sophistication worthy of keeping our niche in the bigger picture of a vibrant water planet.

CHAPTER 4

The Paradigmatic Transformation from Capitalism to Hydroism

Rethinking our species' relationship to the hydrosphere is no longer an academic exercise but a life-and-death concern. This dialogue is more poignant now than ever as our usurpation, manipulation, commodification, and propertization of the waters over the past six thousand years has further compromised life in the past two hundred years by its intimate entanglement with the fossil fuel-based industrial civilization – what's referred to as the water–energy–food nexus. This nexus is not only the indispensable mechanism that powers the industrial society but also the very engine of the capitalist system.

The Water–Energy–Food Nexus

Let's look at the United States as a typical example of how the water–energy–food nexus operates. Ninety percent of the electricity in the United States is generated by thermoelectric power plants, which require the withdrawal of vast amounts of water to generate electricity. Here's how the process works:

> Thermal power plants generate around 80 per cent of the electricity produced in the world, by converting heat into power in the form of electricity. Most of them heat water to transform it into steam, which spins the turbines that produce electricity. After pass-

> ing through the turbine, the steam is cooled down and condensed to start the cycle again, closing the so-called steam cycle. The water is heated with different energy sources (coal, oil, natural gas, uranium, solar energy, biomass, geothermal energy) depending on the sub-type of power plant, but the principle is the same. All power plants need to cool down the steam and most of them use water to do so, which requires them to be near a water source (river, lake, or ocean). Although power plants require water for several processes (steam cycle, ash handling, flue gas desulfurization systems, among others) most of the water requirements – usually about 90% of the total – are for cooling purposes.[1]

In the United States, thermoelectric power accounts for a significant percentage of total water withdrawals, 72% of which are from freshwater sources.[2] Globally, 29% of nuclear power plants use fresh waters from their country to cool nuclear reactors and, in France, where nuclear power makes up nearly 68% of the electricity, the power plants use even greater quantities of fresh water, drawn from the environment yearly.[3] And the waters – especially in places like southern France but also elsewhere – are often so hot in the summer months because of climate change that they cannot be drawn to cool the plants, which are then forced to power down or often shut off operations entirely. Elderly citizens have died over the years in France because of a lack of air-conditioning resulting from nuclear power plants being powered down or shut off in the extreme summer heat. Moreover, the cooled waters that are used in nuclear reactors when cycled back into the environment may further warm ecosystems and undermine agricultural yields.[4]

Both water and energy are also critical for producing food. Here again, a problem arises with positive runaway feedback loops that negatively impact all three of the nexus elements, affecting the availability and resilience of the entire water–energy–food nexus. About 70% of global water withdrawals are used for agriculture, and much of the withdrawals are wasted because of poorly functioning and outdated infrastructure and mismanagement.[5]

Fossil fuel energies are not only used to heat up water to create steam, which spins the turbines that produce electricity but are also used to grow food and fiber-based crops and power farm machinery. The green revolution in agriculture that swept across the world from the late 1950s onwards relied on the widespread use of petrochemical fertilizers and pesticides to grow hybrid high-yield crop varieties. Incredibly, 40% of all the energy used in the global food system goes to the production of artificial petrochemical-based fertilizers and pesticides, which seep into the soil and underground water, creating a toxic brew.[6]

On the water front, every neighborhood and community will need to scale up water harvesting when the rainfall comes where they live and work, store it, and share it when needed. On the energy front, a quick expedited phasing out of fossil fuels and nuclear power in the generation of electricity ought to take precedence in the transition agenda, along with the speedy deployment of solar, wind, geothermal, tidal, and wave energies, and accompanying green hydrogen storage. On the food front, there will need to be a downshift from petrochemical and biotech agriculture and the uplift of ecologically based regenerative agriculture.

The water–energy–food nexus is now taking our species and fellow creatures to the edge. Breaking the nexus is the highest priority, given the extreme shortage of fresh water at various times of the year in the wake of global warming of the environment, with its accompanying droughts, heatwaves, and wildfires, each increasing in intensity and spreading quickly over the planet. The nexus, which has become a societal chokehold, is imploding alongside the geopolitics of the Industrial Age, with far-reaching implications for the reorganization of society.

Solar and wind-generated power are now the cheapest energies in the world, with their fixed costs plummeting and their marginal costs near zero. The new energies are moving online rapidly on every continent. A study conducted by researchers at LUT University Lappeenranta, Finland in 2019 in the journal *Nature Energy* reported that photovoltaic technology only consumes between two and 15% of the water used by nuclear and coal-fired powerplants, and wind turbines only consume between 0.1 and 14% of the water used in nuclear and coal-fired powerplants.[7] The study collected data from 13,863 thermoelectric power plants representing over 95% of the world's thermogeneration power plants. Their study found that the swift transition from fossil fuels and nuclear-generated power to solar and wind-generated power would, in a best policies scenario, reduce water consumption "from conventional power generation by 95%."[8] The uncoupling of the waters from fossil fuel and nuclear-power generation is an inflection point in the freeing of the planet's hydrological cycle.

The deconstruction of the water/fossil fuel/nuclear power nexus by the ever-quickening market penetration of solar and wind generating power is a gamechanger. The fossil fuel-based industrial era moved in lockstep with the rise of global geopolitics as nations fought to control water resources, fossil fuels, and uranium for power generation. In 2015, the top four nations in both fresh water and overall water consumption were China, the U.S., India, and Russia. All four of these leading nations are shored up by huge military establishments wedded to the geopolitics of a fossil fuel-based economic order. The breaking of the water/fossil fuel/nuclear complex and the switchover to a shared solar and wind powered society takes the world from a geopolitical era to a biosphere age, where the sun shines everywhere and the wind blows everywhere, allowing the human race to generate solar and wind electricity collectively where they live and work and share these energies across regions, continents, oceans, and time zones on a now scaling global energy internet.

The capturing of the hydrosphere and lithosphere to power a fossil fuel-based industrial civilization and a water/energy/food nexus would not have been possible were it not for a fundamental change in the way humanity conceived of its relationship to the land. It was the English philosopher John Locke who introduced the philosophical rationale for the privatization of the lithosphere, later picked up by others, extending the propertization argument to the hydrosphere. His thesis on the nature and role of private property provided the intellectual rationale for the development of capitalism.

In the long stretch of the feudal era and the early Middle Ages, property was defined very differently from the way we think of it today. The Church made clear that while the Earth is God's creation, the Lord divvied up the various parts of his domain to his flock along a descending hierarchy of responsibilities and obligations, reaching down from God's emissaries in the Church and, from there, to the kings, nobles, knights, and serfs. In feudal Europe, it was proprietary relations and not property relations that determined how God's creation was shared. Nobody owned property as we think of the concept today but, rather, only stewarded that part of the Lord's creation entrusted to each segment of the flock. The very idea of buying and selling land was more of an anomaly than a mainstay.

By the twelfth century, the feudal order based on proprietary relations was weakening, with the introduction of enclosures, allowing the lords to privatize their lands and sell them in what became a fledgling real estate market, all of which began to lay the foundation for the modern concept of private property relations in a market economy. It was the English philosopher, John Locke, who came to the fore with a philosophical justification for reorganizing society around private property in the publication of his *Two Treatises on Civil Government* in 1690, arguing that private property ought to be considered an unalienable natural right. Locke justified his premise by referring back to the Lord's bequeathing of his creation to Adam and all of Adam's descendants who shall have dominion over life on Earth. Locke wrote:

> When [God] gave the world in common to all mankind, [He] commanded man also to labour, and the penury of his condition required it of him. God and his reason commanded him [humankind] to subdue the earth – i.e., improve it for the benefit of life, and therein lay out something upon it that was his own, his labour. He that, in obedience to this command of God, subdued, tilled, and sowed any part of it, thereby annexed to it something that was his property, which another had no title to, nor could without injury take from him.[9]

While disturbing, Locke took the argument around the possession of property a step further, arguing that nature itself lies in waste until harnessed and transformed into valuable property by the labor of man:

> He who appropriates land to himself by his labour, does not lessen, but increases the common stock of mankind. For the provisions serving to the support of human life, produced by one acre of enclosed and cultivated land, are . . . ten times more than those which are yielded by an acre of land of an equal richness lying waste in common . . . it is labour then, which puts the greatest part of value upon land, without which it would scarcely be worth anything.[10]

Locke undermined the very idea of the Earth as a commons and a shared obligation of each succeeding generation to steward and substituted the utilitarian notion that each individual has an inalienable right to transform the Earth from a life force to a resource for individual gain in the marketplace.

Locke's recalibration of the planet as passive resources waiting to be transformed into private property and wealth that can be bought and sold in the marketplace would be picked up by Adam Smith and legions of economists that followed, making utilitarian self-interest the underlying theme of the Age of Progress. Smith declared that:

> Every individual is continually exerting himself to find out the most advantageous employment for whatever capital he can command. It is his own advantage, indeed, and not that of the society, which he has in view. But the study of his own advantage naturally, or rather necessarily, leads him to prefer that employment which is most advantageous to the society . . . He intends only his own gain, and he is in this, as in many other cases, led by an invisible hand to promote an end which was no part of his intention . . . By pursuing his own interest, he frequently promotes that of the society more effectually than when he really intends to promote it.

Locke's obsession with transforming, privatizing, and commodifying the Earth as the means to generate wealth blinded him and his contemporaries to the fact that the real wealth that animates the life force on Earth is the waters. Without water, photosynthesis would be impossible. Here's what Locke missed:

> During photosynthesis, plants take in carbon dioxide (CO_2) and water (H_2O) from the air and soil. Within the plant cell, the water is oxidized, meaning it loses electrons, while the carbon dioxide is reduced, meaning it gains electrons. This transforms the water into oxygen and the carbon dioxide into glucose. The plant then releases the oxygen back into the air, and stores energy within the glucose molecules.[11]

But photosynthesis is not possible without nature's base capital, soil. No soil, no vegetation, no photosynthesis. The soil is a highly complex microenvironment. Its parental material is rock. Over a long period of time, rock is subjected to physical weathering and natural erosion by the waters and disintegrates into ever-smaller particles, which eventually become sand and sediment. Lichens mix with the sand and sediment, breaking it down to even smaller particles. Fungi and bacteria, burrowing insects, and animals also assist in the degrading of the rock into soil. The minerals in the degraded rock are the baseline ingredients of the soil. Plants, in turn, grow in the soil. Animals eat the plants and contribute their feces to the soil. Worms and bacteria break down plant litter and animal waste, adding to the soil base. The average soil sample is composed of 45% minerals, 25% water, 25% air, and 5% organic matter. There are more than seventy thousand types of soil in the United States alone.[12] The minerals found in the degraded soil work their way up the food chain, from plants to animals and to our own bodies.

The minerals extracted from the rocks found in human bodies include calcium, phosphorus, sodium, potassium, sulfur, chloride, and magnesium.[13] And the soil, which is the foundation of the lithosphere, is made possible by the forces of the waters. But the waters also play a crucial role in making possible the atmosphere, the other great sphere on Earth. Ocean phytoplankton produce 50% of the oxygen on Earth.[14] Scientists believe that oxygen was largely absent from the atmosphere for the first two billion years. "But at some point, Earth underwent what scientists call the Great Oxidation Event . . . as ocean microbes evolved to produce oxygen via photosynthesis."[15] The oxygen rose into the atmosphere, allowing life to evolve on land and water.

And now, after capturing, commodifying, propertizing, and consuming the Earth's four great agencies – the hydrosphere, lithosphere, atmosphere, and biosphere – that regulate life, the planet's spheres are revolting beyond our feeble attempts to hold them at bay or even comprehend . . . and it's the hydrosphere that is defining the new rules that will determine the future evolution or devolution of life on Earth.

As our species begins to fully grasp the role that the waters of the hydrosphere play in the evolution of the other underlying agencies of the Earth, Locke and Smith's naive notion of capital as "man-made" things that confer value (which generally include machinery, intellectual property, and financial assets that facilitate the production, exchange, and consumption of goods and services and the accumulation of wealth) is completely misconceived. Rather, the hydrosphere is the animating force on our self-organizing and ever-evolving Earth, and the indispensable medium upon which all life thrives.

If there is a single marker in how the capitalist system is attempting to negotiate the new understanding of the primary role that the waters play in affecting the global economy, it would have to be how world markets address the utilization of the waters in agricultural production, commerce, and trade. That story has come to light in recent decades, with the introduction of a new term called "virtual water." The term was coined back in 1993 by Tony Allan, then a professor of geography at King's College London. His tale on why he invented the term and the subsequent discovery of the dark side of what the new calculus exposed about the way water is used is eye-opening. His data show how the capitalist system, in tandem with governments, has conspired with global agro-business companies to drain the remaining freshwater reserves of the planet, while pretending to represent sustainable water practices.

Allan said he had invented the term – earlier he had referred to it as "embedded water" – with the intent of explaining "why our unsustainable political economies of food–water resources exist."[16] Virtual water is a way of calculating the total amount of water that is used in producing food, fiber, and energy. Virtual water is derived from what's known as the "water footprint," which is the calculus of the water required for producing goods and services consumed by the inhabitants of any given country. The water footprint is further divided into the volume of domestic water used in producing the goods and services versus the imported water used in other countries to produce the goods and services exported to the user country.[17]

A bag of potato chips requires 48.9 gallons of virtual water. A cup of coffee requires 140 liters of virtual water to grow, package, and ship the beans. A pound of butter requires 3,602 virtual gallons of water.[18] To fully grasp the scope and extent to which water defines our dietary habits, consider the fact that a single almond requires 3.2 gallons of water to produce. The principal almond-producing region is in the Central Valley of California and accounts for 80% of all the almonds produced on the planet. Add it all up, and ten percent of all the water consumed by agriculture in California annually goes to quenching the thirst of the almond trees in the Central Valley – that's more water than is consumed by the entire populations of Los Angeles and San Francisco in a year.[19]

Every nation's water footprint is continually updated. India, for example, makes up 17% of the global population and contributes 13% of the global water footprint. The U.S., by contrast, has the largest "per capita" water footprint.[20] The calculus of virtual water is often used to compare the volume of water in producing one commodity versus another. For instance, thirteen hundred tons of water are consumed to produce a ton of wheat, while sixteen thousand tons of water are used to produce a ton of beef.[21] The wheat/beef comparison becomes particularly interesting when we realize that upwards of 50% of the dietary dry matter that goes to cows is in the form of wheat.[22]

When evaluating the virtual water calculus in terms of daily food habits, a heavy beef-eater consumes upwards of five cubic meters of virtual water daily, compared to a vegetarian, who consumes only 2.5 cubic meters of virtual water.[23] Although most of us assume that the vast amount of fresh water consumed each year goes to human consumption – drinking, bathing, cooking, cleaning, and the like – in actuality, agricultural production accounts for a whopping 92% of global water consumption, while industry uses 4.4% of the fresh water, and household water consumption comes in dead last, using only 3.6% of all the fresh water consumed each year.[24]

Because agriculture is far and away the largest consumer of water, it's best to start there. When it comes to virtual water, we're talking about green water – rainfall – which is stored in soil and can't be pumped but is taken up by plants and trees, and blue water in rivers, streams, lakes, and reservoirs, which can be pumped into pipes and sent to users. Allan makes the distinction that while water taken up by plants and consumed by farm animals can be green or blue, drinking water is always blue. Green water makes up "80% of the virtual water embodied in the 20% of food that is traded internationally."[25]

To decipher how the calculus of virtual water affects agriculture and the remaining available fresh water on Earth, it's helpful to identify the players – that is, who are the handlers and suppliers, and who are the recipients. The global food trade is frighteningly lopsided. Ten exporting countries – the U.S., China, India, Brazil, Australia, Argentina, Canada, Indonesia, France, and Germany – are the principal exporters of virtual water via agricultural trade, while other countries are virtual water importers.[26] The latter are nations whose food-production capacity is insufficient to feed their population generally because their topography is arid or semi-arid, and their green waters are diminishing with the warming of the climate. Then, too, their historically low freshwater reserves are rapidly declining as they desperately search for enough available water to provide for drinking and bathing and the creation of steam to run turbines and generate electricity for operating factories.

The skewed ratio between the have and have-not nations relative to freshwater availability is even more straitjacketed by the fact that global agribusiness companies – ADM, Cargill, Louis Dreyfus Company, Bunge – all U.S. and French conglomerates – dictate much of global agriculture and trade, and only ten water companies – Eversource Energy, Acwa Power Company, American Water Works Company, XYLEM, Hong Kong & China Gas Company, Suez SA, Essential Utilities, United Utilities, Severn Trent, and Companhia de Saneamento Básico do Estado de São Paulo – control much of the global commerce and trade in water.[27]

The virtual water calculus becomes particularly dicey and controversial when it comes to the global trade in agricultural products. Importing countries facing an ever-diminishing supply of fresh water are reluctant to grow food if it means running short of water for human consumption and sanitation and prefer to import food from a handful of major exporting nations – in a very real sense, importing "virtual water" to feed their people: what we're talking about here is the actual water used in the exporting country that went into the foods being shipped to the importing countries. Practically speaking, the imported food is cheaper than the cost of producing the food domestically and saves precious water for vital human services.

Here is where the sleight of hand comes in. The reason the handful of exporting nations can sell cheaper food to the 160 water-starved countries is that the farms and global agribusiness companies are heavily subsidized by their governments, allowing them to prosper in the global marketplace, while diminishing their own countries' remaining freshwater reserves. Take the United States. Although it's doubtful that the majority of American voters are knowledgeable of the game underway, the fact is that the U.S. government underwrites the entire farm sector on a scale that would shock consumers. In 2020 alone, U.S. government aid to farms accounted for an incredible 39% of net farm income, or 46.5 billion dollars.[28]

This extraordinary subsidy – or giveaway – allows farmers to pump ever-increasing amounts of stored water from once bountiful U.S. aquifers and reservoirs, leaving them nearly bone dry and threatening the water security of the country, while global agribusinesses and the farmers that serve them sell cheap food commodities to water-starved countries everywhere. What this means is that "food exporting economies charge neither for water consumed in food production nor for the costs of damaging their water ecosystems and their natural biodiversity. The biggest attraction, however, of importing food staples such as wheat has been that the food commodity prices enjoyed by importers of staple grains did not even reflect the full costs of farm production."[29] Prices were subsidized by the exporting nations.

And now, for example, in the United States the entropy bill has come due and it signals the potential demise of groundwater across the entire continental landmass, threatening the very viability of the wealthiest country in the world. On August 28, 2023, a *New York Times* investigative team released a massive study on the current state of the continent's groundwater resources. The investigators monitored 84,544 wells since 1920 and found that half the sites "have declined significantly over the past 40 years as more water has been pumped out than nature can replenish."[30] In just the past decade, four out of every ten sites have "hit all-time lows." *The New York Times* goes on to report that:

> Many of the aquifers that supply 90% of the nation's water systems, and which have transformed vast stretches of America into some of the world's most bountiful farmland are being severely depleted. These declines are threatening irreversible harm to the American economy and society as a whole.[31]

The groundwater being pumped is used by farms around the country to irrigate crops at little cost, allowing a subsidy of sorts – and, more importantly, a massive handout to giant agri-businesses, who then sell the food to less water-endowed countries, while depleting America's remaining groundwater reserves.

The *New York Times* investigative team warns that:

> America's lifegiving resource is being exhausted in much of the country, and in many cases won't come back. Huge industrial farms and sprawling cities are draining aquifers that could take centuries or millenniums to replenish themselves if they recover at all.[32]

These partially hidden government subsidies by the major exporting nations keep the farmers happy and the global agribusiness companies awash in revenue. A new study by the World Wildlife Fund estimates the total quantifiable use value of water at $58 trillion, equivalent to

60% of global GDP in 2021.[33] And, as previously mentioned, 70% of global freshwater withdrawals goes to agriculture. American consumers, in turn, remain relatively sanguine in their grocery shopping – at least until the recent inflationary spiral caused in part by the war in Ukraine and oil-price hikes. What's lost in the subterfuge is that America and the other exporting nations are already facing down on ever-depleting water reserves in a fast-approaching showdown, when the deceit comes home to roost.

Although no one would deny the value of measuring the water footprint and the virtual water embedded in every economic activity, there is shamefully little discussion about the need to transform the food diet in both exporting nations and importing countries away from foods with a high virtual water calculus – especially beef related but also other meat-based food sources – to an all-encompassing plant-based food culture. There are scores of drought-tolerant vegetables with high protein and vitamin added value that thrive with little need of water, including asparagus, rhubarb, pole beans, Swiss chard, squash, arugula, and okra.[34] Drought-tolerant fruits include apples, pomegranate, mulberry, grapes, figs, blackberry, apricots, and plums.[35]

Reading between the lines, it's obvious that the major virtual water exporting countries – even Brazil, whose hydropower has shut down periodically because of a shortage of water due to climate-induced drought in the Amazon – are fast depleting their waters and will need to make a U-turn and transition away from beef and other meats and the production of grains to feed livestock. Currently, 55% of the world's crop calories feed people directly, while 36% of the food is used for livestock and 9% for biofuels.[36] Perhaps the real virtue in using a virtual water calculus to measure the waters used in provisioning various foods is that it will encourage a change in dietary habits, taking our species off a meat-based dietary regime and onto a largely vegetable and fruit-based diet of higher nutritional value, while ensuring better health and preserving the waters – especially so if that information were to be included in the labeling of all food products.[37]

Leaving Capitalism Behind

Our new understanding of the life force on Planet Aqua is taking us from capitalism to hydroism as we journey further into the next stage of the Earth's evolution. Capitalism champions growth, while hydroism bolsters flourishing. Capitalism pursues productivity while hydroism spurs regenerativity. Capitalism regards nature as "passive resources." Hydroism, in contradistinction, views nature as animated "life sources." Capitalism generates negative externalities while hydroism fosters circularity. Capitalism measures economic success by gross domestic product (GDP) while hydroism measures happiness by quality-of-life indicators (QLI). Capitalism prioritizes globalization while hydroism seeks glocalization. Capitalism pursues geopolitics while hydroism focuses on biosphere politics. Capitalism is intimately enjoined to nation-state sovereignty, while hydroism favors a partial extension to bioregional governance. Capitalism thrives in markets, whereas hydroism prospers in networks. Capitalism engages in zero-sum games. Hydroism, in stark contrast, sparks the network effect. Capitalism is powered by fossil fuels and nuclear energy whereas hydroism relies on near-zero marginal cost solar and wind energy to power the economy.

The deeper question is what has led to such diametrically opposed ways of conceiving and organizing economic activity? It's because capitalism is modeled after Newtonian physics, while hydroism draws on the laws of thermodynamics, which changes the very way we think about economics. Neoclassical and neoliberal economic theory and practice draw their operating assumptions from the atemporal nature of Newton's laws and largely dismiss the temporal nature of all economic activity whose externalities ripple far out into the future, leaving their imprint, regardless of how small, on the patterns, processes, and flows of the Earth's dynamic spheres – the hydrosphere, lithosphere, atmosphere, and biosphere. A new generation of ecological economists, however, base all their economic assumptions and deliverables on following the path set out by the first and second laws of thermodynamics and the entropy trail that accompanies each exchange with nature affecting every future event yet to come on Planet Aqua.

Capitalism takes a rational, detached, and utilitarian approach to scientific inquiry, while hydroism studies all natural phenomena with a biophilic attachment using complex adaptive social/ecological systems modeling (CASES) to understand the relationships and synergies between the hydrosphere, lithosphere, atmosphere, and biosphere and the evolution of life across the natural world.

CASES engages society in a wholly new scientific realm that is at odds with the traditional Baconian approach to scientific exploration, with its deep relationship to capitalism and the Age of Progress. Francis Bacon, considered the founding father of modern science, whose view on scientific inquiry became the blueprint of the Enlightenment, introduced inductive reasoning to pry into nature's secrets, for the exclusive purpose of increasing the opulence of our own species. Baconian science pursues objectivity, detachment, prediction, and pre-emption of nature as the *sine qua non* of human agency – a utilitarian liturgy that would become the go-to academic playground for "managing" the natural world in the modern era and a helpmate to capitalist theory and practice.

By contrast, complex adaptive social/ecological systems approach nature as "open dynamical systems that continuously self-organize their structural configuration through the exchange of information and energy."[38] If there is a defining approach to the new science, it's that complex adaptive systems learn to be responsive to each new process, pattern, and flow in an animated and ever-evolving planet – known as emergence.

CASES scientific inquiry falls short of the kind of predictability that science has hitherto sought, and for good reason. Attempts to establish boundary lines on the animated and evolving spheres of the planet completely misses the point that all self-organizing systems are patterns among other patterns, which spread out in time and space and across the earth's operating spheres, impacting one another in subtle and profound ways that can rarely be foreseen. The profound lesson when applying CASES thinking is to partially let go of the obsession with "prediction" and "pre-emption," and prioritize "anticipation" and "adaptation."

Interestingly, today's transformative shift in scientific inquiry was anticipated by John Dewey, Charles Sanders Peirce, William James, and George Herbert Mead, the pioneers of pragmatism in science in the late nineteenth and early twentieth centuries. They took umbrage with the inductive and deductive framing of scientific engagement that fetishized pre-emption and utilitarianism at the expense of nature's bounty. The pragmatist scientists introduced what they called "abductive scientific inquiry," which prioritizes deep mindful participation with nature – that is, listening to and anticipating nature's callings and adapting to the moment-to-moment evolution of the hydrosphere and accompanying lithosphere, atmosphere, and biosphere. Their science is one of continuous integration with an animated Earth and not seizing, pacifying, and pre-empting the evolution of the natural world.

Complex adaptive social/ecological systems modeling is asymmetrical to a capitalist approach to engaging nature. With CASES, "management" of nature gives way to "stewardship" of nature, and the hydrosphere becomes the primary agent whose complex interactions with the lithosphere, atmosphere, and biosphere determine the evolutionary well-being of our water planet and ultimately our economic life. This is the critical inflection point that takes our species from capitalism to hydroism.

CASES has evolved from earlier incarnations of complexity theory, which emerged in the second half of the twentieth century across a number of academic disciplines. Its spearheads included the Nobel Laureate in chemistry, Ilya Prigogine, whose study of dissipative structures in non-equilibrium thermodynamics helped introduce complexity theory, and Edward Norton Lorenz, best known for his study of weather systems and non-linear causal pathways, which was popularized by the term "the butterfly effect."

While complexity theory in science is new, earlier generations of our species far back in the historical record and well before the Enlightenment, the Age of Progress, and modern capitalism engaged in more simplified approaches to complexity theory and practice in the stewardship of commons across shared ecosystems.

As already mentioned, Elinor Ostrom reintroduced the long-buried history of the Commons and, among others, revived traditional complexity practices used over eons of history in the management of ecosystems. Her work and others' on polycentric systems are among the several tributaries of the emerging science of complex adaptive social ecological systems modeling that are transforming all of our traditional notions of scientific inquiry and practice.

How different the commons approach to economic activity is to the capitalist approach. The latter doesn't account for the voluminous negative externalities that trail the short-term market gains of economic activity – leaving future generations for millennia to come with an impoverished Earth. When it comes to settling the entropy bill brought on by global warming emissions, the figures are mindboggling. In a paper published in September 2023 as a preprint by the National Bureau of Economic Research, a group of distinguished scientists and economists estimate that "by 2100, the cost of damages associated only with past U.S. emissions could rocket past $100 trillion, with future emissions only adding to the total."[39]

This emerging struggle between capitalism and hydroism is already coming to the fore. But, instead of meeting the new reality head-on and adapting to a fast-transforming hydrological cycle that's tearing down all of the standard markers by which we have come to organize our economic life, the capitalist system is taking a rearguard approach in attempting to salvage its remaining capital investments. The strategy, already gaining traction, is lopping off the accumulative negative externalities brought on by climate change from the market – in a sense, cutting their losses, which is only narrowing the market's playing field.

Consider a new term gaining momentum called bluelining, reminiscent of the old *redlining* of neighborhoods by banks, insurance companies, and local governing jurisdictions demarcating mostly poor neighborhoods populated by black Americans and other people of color as a means of denying lending and investment services and especially home mortgages. Bluelining is suddenly occurring across the United States as banks and insurance companies demarcate high-risk climate disaster regions to deny insurance to homeowners and local businesses in these communities. Déjà vu, once again, as bluelining is occurring in mostly low-income neighborhoods populated by blacks, Hispanics, and other peoples of color.[40]

Still, even as millions of Americans are experiencing greater destruction of property and lives, the unraveling of ecosystems, and the extinction of wildlife brought on by the water–energy nexus, inertia in governing codes, regulations, and standards stands in the way of freeing the waters and letting ecosystems rewild and evolve in new life-affirming ways. Bear in mind the fact that the National Flood Insurance Program (NFIP) of the United States government generally only allows flooded buildings on floodplains and near coastal waters to be repaired and rebuilt onsite, meaning buildings are repeatedly flooded and repaired by the dollars of American taxpayers under NFIP. Between 1989 and 2018, multiple insurance payments of more than $22 billion were given out to 229,000 "repetitive flood loss properties."[41]

National flood insurance given to continually rebuilding in the same high-risk disaster-prone regions makes absolutely no sense when studies project that upward of 162 million Americans, one half of the population, will likely experience a deterioration of their environment, as climate-related disasters continue to evolve and accelerate in their communities, and 93 million Americans will face "severe" climate-related impacts.[42] Bluelining is the way a capital-directed society measures risk and generates so-called wealth, while a commons approach follows the lead of the waters on its rewilding journey, continually adapting to its planetary reorientation while finding new ways to flourish.

For all the talk about addressing climate change, governments, the marketplace, and civil society have yet to come to grips with what it means that the waters are freeing themselves in regions everywhere. To cite a couple of examples of how the water cycle is transforming itself on a global warming planet, consider what occurred in the city of Jackson, the state capital of Mississippi, over the late summer of 2022. Torrential rains poured down on the city and its suburbs, resulting in mass flooding, which crippled the city's aging water plant and hydro-infrastructure, leaving the entire system in shambles and stranding its 150,000 residents without access to water.[43] City officials said it would be months or even years before a fully functioning water system can come back online. Weeks earlier, 25,000 people in eastern Kentucky were hit by deadly floods decimating communities and breaking apart antiquated water lines with no relief in sight.[44] Concurrently, Texas water utilities were caught unprepared when towns and cities across the state were struck by record heatwaves and droughts, and the hardening of soil, breaking apart water mains. And this followed an earlier winter storm that froze water pipes, causing thousands of them to burst.[45]

A city-council member in Jackson summed up the problem facing his community and communities around the world, lamenting that "our flood controls, our systems that are in place, are extremely antiquated."[46] This quip is being heard in thousands of communities. Mikhail V. Chester, a professor of civil, environmental, and sustainable engineering at Arizona State University, struck an ominous note that our species has yet to come to terms with – the fact that "the climate is simply changing too fast relative to how we can change our infrastructure."[47]

But it's not only the water infrastructure that is collapsing all over the world. Every other part of the aging twentieth-century infrastructure in world-class cities, suburbs, small towns, and rural hamlets is falling prey to a hydrological cycle caught up in climate change and freeing its waters – and ripping up the entire infrastructure of the modern age. And, it's not just the scale, but the ubiquity of the impact that a rampant hydrosphere is having at the granular level of where we work and live that makes the present situation so frightening to comprehend.

Take the opening days of classes in the last week of August 2022 in Philadelphia's public schools in the nation's sixth largest city. Tens of thousands of students were sent home shortly after arriving for classes because of lack of air-conditioning during a heatwave.[48] More than 60% of the school buildings in the Philadelphia school system are without sufficient air-conditioning and many other buildings are only partially air-conditioned.[49] The problem is that the average age of the buildings is 75 years, and their electrical system can't support central air in most of the structures.[50] City officials report that the problem is magnified by billions of dollars in unmet capital needs. Even ensuring that this hurdle can be jumped, the magnitude of the task of overhauling the city's aging infrastructure might not be complete until 2027 at the earliest.[51] But, here again, catching up to a changing climate – in this instance preparing schools for 90-degree weather – may fall short of the mark, as climate change records normal temperatures of 95 degrees or even 105 degrees in the years and decades ahead. Not able to catch up to an ever-rising change in temperature might force the Philadelphia school system and other school systems to shut down for longer days, or even weeks, in the early fall semester, and end the school year in early June or even late May, as the temperatures rise once again, further reducing the school year and the education of students.

Yet, few if any public officials in America and elsewhere are even hinting that entire cities, suburbs, and rural communities caught up in frigid winter temperatures and record-breaking blizzards, thousand-year-floods in the spring, long droughts and excruciating heatwaves and wildfires in the summer, and powerful category three, four, and five hurricanes in the fall, might need to rethink how they live and where they live in the decades ahead. But what nobody wants to talk about is that hundreds of millions and even billions of people in the coming decades will need to migrate to more temperate regions and, even then, likely temporarily, as a dwindling global population searches for remaining safe havens.

This is what humanity is confronting and needs to own. The waters are freeing themselves and our species will have to learn to adapt to the waters and let go of the naive fiction that we can continue to adapt the water to our species' whims. This wake-up call will require a change in how we position ourselves in time and space and on a scale only experienced once before in the history of our species, when we transitioned from the last Ice Age of the Pleistocene to the temperate climate of the past 11,000 years.

PART II

The Canary in the Mine

How the Mediterranean Eco-Region Became Day Zero on a Warming Earth and a Bellwether of the Second Coming of Life

CHAPTER 5

The Near Death and Rebirth of the Mediterranean

If there is anything like a discernable trajectory to our species' historical journey over the past six millennia of history, it would have to fold into the rise and fall of hydraulic civilizations. Hydraulic civilizations follow along a classical bell curve. As mentioned earlier, the cradle of civilization dates back to six thousand years ago in the foothills of ancient Anatolia – present-day Turkey – at the headwaters of the powerful Euphrates and Tigris rivers. These were the first great rivers of the world to be shackled, sequestered, modified, and exploited to flood and irrigate fertile lands and produce a cornucopia of surplus grains – mostly wheat and barley – in a region including present-day Turkey, Syria, Iraq, and Iran. Everything we identify today as the makings of civilization began along these rivers and it's these rivers that are now drying up in the lifetime of today's babies, over the course of the next several decades. It's here that the first scaled urban hydraulic civilization arose, laying the foundation for the evolution of successive waves of hydraulic civilization and the rise of our species as the dominant species on the planet and now to a mass extinction of life on Earth.

This first record of urban civilization was given the name Mesopotamia, a Greek word that means "the land between the rivers."[1] It was here that villages first turned into cities with populations reaching upwards of 40,000 to 50,000 inhabitants.[2] Mesopotamia spawned several of the early great civilizations, including the Sumerian, Akkadian, Babylonian, and Assyrian empires and supported upwards of twenty million people over a period of time.[3]

These hydraulic civilizations, and others that followed in Egypt, the Indus Valley in India, the Yellow River Valley in China, Crete, Greece, the Roman Empire, the Khmer civilization in Southeast Asia, the Mayan civilization in what is now southern Mexico, Guatemala, and northern Belize as well as the Inca Empire in western South America experienced similar bell curves. Some declined because of changes in the climate and others because of invasions by marauding tribes, and still others because of a mounting entropic bill. Although anthropologists and historians have devoted considerable time to the first two causal agents, far less time has been given over to the entropic debt that inevitably rode side-by-side with hydraulic infrastructure.

The hydraulic empires of the Mediterranean, India, and China gave rise to a great leap forward in human consciousness and the first bloom of cosmopolitanism. But, in the end, they were unable to escape the verity of the Second Law of Thermodynamics. A strong body of research into the rise and fall of hydraulic civilizations has shown that while there are many explanations that account for their eventual demise, at the very top of the list is the entropy bill brought on by the changes in soil salinity and sedimentation.

In Mesopotamia, the alluvial soils were carried inland by river and irrigation water. The irrigation waters contain calcium, magnesium, and sodium. As the water evaporated, the calcium and magnesium precipitated as carbonates, leaving the sodium embedded in the earth. If not flushed down into the water tables, the sodium ions are absorbed by colloidal clay particles and the soil becomes impermeable to water. Heavy concentrations of salt stall the germination process and impede the absorption of water and nutrients by plants.[4]

Southern Iraq, for example, experienced a serious problem with salinity between 2400 BCE and 1700 BCE, while central Iraq experienced a similar crisis between 1300 BCE and 900 BCE. An increase in the salinity of the soil forced a shift from wheat cultivation to the more salt-tolerant barley crop. In 3500 BCE the proportion of wheat and barley production was roughly equal. However, in less than a thousand years, the less salt-tolerant wheat crop made up less than one-sixth of the agricultural output. By 2100 BCE, wheat made up only two percent of the crops in the same regions and, by 1700 BCE, wheat cultivation had ceased in the southern alluvial plain.[5]

The salinization of the soil also led to a decline in fertility. In the city of Girsu, for example, the average agricultural yield was 2,537 liters per hectare in 2400 BCE but declined to 1,460 liters per hectare by 2100 BCE. In the neighboring city of Larsa, the yield had slipped to only 897 liters per hectare by 1700 BCE.[6] The impact on the cities whose populations depended on agricultural surpluses to maintain an urban way of life, was devastating. Sumer's city-states became embroiled in political and economic turmoil leading to the disrepair and even collapse of much of their complex infrastructures, while their populations plummeted.[7] The very hydraulic infrastructure that provisioned an enormous increase in the flow of water allowing the Sumerian population to erect the first great urban civilization, establish a primitive cosmopolitanism, and advance the notion of individual selfhood, led to an equally fateful entropic impact on the region's ecosystems canceling out many of the gains, and leaving this early experiment in establishing a hydraulic urban civilization and its surrounding environment impoverished. Thorkild Jacobsen and Robert M. Adams of the Oriental Institute of the University of Chicago, in their study on the subject, published in the journal *Science* more than a half-century ago, concluded that:

> [p]robably there is no historical event of this magnitude for which a single explanation is adequate, but that growing soil salinity played an important part in the breakup of Sumerian civilization seems beyond question.[8]

The increasing salinity of soil led to massive crop failures and a similar entropy crisis in the Indus Valley 4,000 years ago.[9] Likewise, archaeologists have found evidence of soil salinity leading to catastrophic crop failures and the abandonment of territory in the ancient Mayan hydraulic civilization in Central America.[10] The salinization of soil and entropic buildup have been a driving "factor in the weakening and collapse of complex hydraulic civilizations throughout history, reaffirming the inescapable relationship between increasing energy throughput and a rising entropy debt."[11]

Complex hydraulic civilizations are accompanied by two seemingly disparate phenomena – both an expansion of empathic engagement in ever-more sophisticated urban environments and an increase in the entropy bill. So, both empathy and entropy evolve simultaneously with the evolution of complex hydraulic civilizations, which raises a paradox that's gone unexplored by anthropologists, historians, and philosophers.

I first came across the paradox in a seven-year period between 2003 and 2010 when I turned my attention to the role empathy has played in the historical development of our species. I had written about empathy in several previous books over a thirty-year period, but never in depth. This time, I decided to explore the evolution of empathy in greater detail – its anthropology and history – and its effect on the other salient aspects of society including our family and social life, economy, modes of governance, and worldviews. Somewhere far along into the study, I became aware of the paradox, and I admit it shook me. Here's what I found and wrote in *The Empathic Civilization*:

> At the very core of the human story is the paradoxical relationship between empathy and entropy. Throughout history new energy regimes have converged with new communications revolutions, creating ever more complex societies. More technologically advanced civilizations, in turn, have brought diverse people together, heightened empathic sensitivity, and expanded human consciousness. But these increasingly more complicated milieus require extensive energy use and speed us toward resource depletion.
>
> The irony is that our growing empathic awareness has been made possible by an ever-greater consumption of the Earth's energy and other resources, resulting in a dramatic deterioration of the health of the planet.
>
> We now face the haunting prospect of approaching global empathy in a highly energy-intensive, interconnected world, riding on the back of an escalating entropy bill that now threatens catastrophic climate change and our very existence. Resolving the empathy/entropy paradox will likely be the critical test of our species' ability to survive and flourish on Earth in the future. This will necessitate a fundamental rethinking of our philosophical, economic, and social models.[12]

The key to decoupling the empathy/entropy paradox is the "great reset." After six millennia of adapting nature to suit the exclusive needs of our own species, we will need to reverse course and adapt our species to the life-affirming processes and patterns of the Earth and find our way back into nature's fold. That's what biophilia consciousness is all about. What's missing is not the way but the will. Hopefully a younger generation, infused with biophilia consciousness, will pave the way forward. Today, the hydraulic infrastructure is collapsing, not only because of the thermodynamic bill, although the buildup of sodium in the soil making it impermeable to water is a perpetual problem that travels in tandem with all hydraulic infrastructure. It is also the dangerous warming of the planet that is depleting the Earth's rivers, lakes, and streams, leaving behind dry water beds everywhere. The last time the Earth was warmer than now was 125,000 years ago, according to a 2021 report issued by the UN Intergovernmental Panel on Climate Change (IPCC).[13]

The World Meteorological Organization (WMO) *2022 State of Climate Services* report is a cautionary tale of how far down the line we are to a dramatically warming planet.[14] The WMO zeroed in on the impact global warming is having on hydraulic infrastructure everywhere, which is so critical to power generation. They found that drought and heatwaves associated with climate change "are already putting existing energy generation under stress." Here is a summary of WMO's findings:

> In 2020, 87% of global electricity generated from thermal, nuclear, and hydroelectric systems directly depended on water availability. Meanwhile, 33% of the thermal power plants that rely on freshwater availability for cooling are already located in high water stress areas. This is also the case for 15% of existing nuclear power plants, a share expected to increase to 25% in the next 20 years. 11% of hydroelectric capacity is also located in highly water-stressed areas. And approximately 26% of existing hydropower dams and 23% of projected dams are within river basins that currently have a medium to very high risk of water scarcity.[15]

On September 13, 2023, a biblical flood rained down on Libya, collapsing its two Wadi Derna dams in a matter of moments, washing entire communities into the floodwaters and sweeping thousands of individuals to their deaths. The Libyan tragedy is a cautionary marker that's likely to be repeated with increasing intensity in the decades ahead. In India and China alone, which are home to 2.8 billion human beings, there are 28,000 large dams that are more than 75 years old and in disrepair and vulnerable to collapse, threatening the lives of literally millions of people. Lest we think U.S. dams are better built and maintained and getting a reprieve – far from it. The average age of dams in the

United States is 65 years or older and it is estimated that "2,200 structures are at high risk of collapse."[16] A rewilding hydrosphere is going to tear apart the entire global hydraulic infrastructure over the next 75 years, putting entire cities and regions at risk of being washed away.

Commenting on the report, the Chief of the WMO warned that "time is not on our side and our climate is changing before our eyes," adding that we need "a complete transformation of the global energy system" given the depletion of lakes and rivers around the world.[17]

Every nation will need to underwrite a formidable transformation in its water–energy–food nexus from fossil fuels and nuclear power to solar, wind and other renewable energies, and from water intensive cash crops – corn, wheat, rice, barley, oats, rye, etc. – for export to water sparing crops – roots and tubers including potatoes, yams, carrots, cassava, and beets – for domestic consumption and export. As well, every country will have to switch from dependence on a diminishing volume of lake water and big river waters to greater reliance on the many highly distributed methods of water harvesting onsite where they live and work, including sponge capture and retainment of rainwater in distributed cisterns and water microgrids along with the resurrection of long buried rivers, streams, aquifers, and wetlands.

The transition from a progress-driven to a resilient-oriented infrastructure is already underway and beginning to scale, affording the Mediterranean nations the opportunity to be among the leaders in advancing the coming Age of Resilience. The Mediterranean ecoregion will likely be the test case that will tell us whether and to what extent our species and fellow creatures will be able to recoup, survive, and thrive on a rewilding planet.[18] A successful transitioning from a dying hydraulic civilization to a resilient ecologically driven society lies in the reconfiguration of the water–energy–food nexus. The features of that reconfiguration are already melding in the Mediterranean and elsewhere.

Day Zero: When Nations Run Dry

What the WMO report is telling us is that much of the hydraulic infrastructure is fast becoming a "stranded asset" as the waters that feed it and power much of our society, irrigate our agricultural fields, and provide drinking water for our species are drying up. And, interestingly, of all the ecoregions in the world caught up in the mayhem of a warming planet and the drying up of the waters, it's the Mediterranean, the cradle of hydraulic civilization, that after six millennia of history is the "canary in the mine" and most at risk of being partially uninhabitable as early as the second half of the twenty-first century, as rivers and lakes run dry, signaling the beginning of the end of the long history of hydraulic civilization in the region. Other hydraulic societies are facing a similar fate. Climate scientists are now zeroing in on the Mediterranean, and for good reasons. What's happening there foretells what's going to unfold in other regions, throwing the whole of humanity into crisis.

To begin with, historically the Mediterranean region has been characterized by dry hot summers with extended periods of drought, and mild to cool wet winters with strong winds and intense precipitation. But now, climate scientists are warning that the millions of people in countries that share the Mediterranean ecosystem are at greater risk than any other populations on the planet with the ramping up of the temperature from global warming emissions.[19] The statistics are frightening. The Mediterranean is warming up 20% faster than the world as a whole.[20] The region's eco-region will have 40% less precipitation in the winter rainy season and, by 2050, 20% less rainfall during the summer season from April through September, with drought conditions prevailing for six months of the year.[21] Water basins are expected to decrease by 25% by 2050.[22] A spate of new studies and reports warn that the Mediterranean region shows "the greatest decline of projected rainfall of any landmass on Earth."[23]

Recent reports have set off alarm bells, especially in Turkey, Syria, Iran, and Iraq, who all share the waters of the Tigris and Euphrates Rivers. Studies show that water amounts have dropped by 40% in the last four decades.[24] The IPCC projects a decline of 29% in the water flow of the Tigris River and 73% in the Euphrates River in the years ahead.[25] Even these projections are quickly being eclipsed with each passing year.

New studies suggest that temperatures in Turkey are likely to exceed 40 degrees Celsius – 104 degrees Fahrenheit – for extended periods of time in the summer months by 2050.[26] And, already, 60% of Turkey's landmass is currently prone to desertification.[27] While climate change is a critical factor, it also must be recognized that Turkey has lost one half of all of its wetlands from development over the last hundred years.[28]

What happens in Turkey doesn't stay in Turkey. Syria, Iraq, Iran, and even Kuwait also share the Tigris–Euphrates River Basin. Iraq alone depends on the Euphrates and Tigris rivers for 98% of its surface water while the country's annual temperature is increasing at "almost double the rate of the Earth's."[29]

Although over the past century the Earth has warmed by 1.3 degrees Celsius, Iraq has warmed by 2.5 degrees Celsius, taking the country far beyond the tipping point. A December 2021 report issued by Iraq's water minister has warned that the Euphrates River could be dry across its terrain by 2040, leaving Iraq without a source of water. Headlines in the Iraqi media warned that "Iraq will be a land without rivers by 2040."[30]

While climate change is the elephant in the room, there are other geopolitical forces at work, particularly the frenzied construction of massive new dams in neighboring Turkey, further curtailing the flow of the Euphrates and Tigris Rivers into Iraq, Syria, and Iran. Turkey is engaged in a Herculean hydraulic infrastructure rollout to sequester the remaining river flows of the Euphrates and Tigris within its political boundaries. At least nineteen new dams have been hastily constructed along the Euphrates and Tigris, and three more dams are on the drawing board.[31]

Meanwhile, an Iraq water ministry report issued in 2020 concluded that within two years Iraq's water supply would not be able to meet demand and by 2035 the water deficit would result in a 20% reduction of food.[32] The United Nations classifies Iraq as "the fifth most vulnerable country in the world" to climate change, and aid groups in the region are warning that "in the short run, millions of Iraqis and Syrians will no longer have access to water, food, and electricity, forcing a refugee crisis of epic proportions."[33]

What's becoming clear is that the Mediterranean region, once the cradle of hydraulic civilization, is collapsing in real time as the Euphrates and Tigris Rivers run dry. An eerie discovery surfaced in 2022 when German and Kurdish archaeologists announced they had uncovered a long-submerged ancient city in Iraq after large amounts of water were withdrawn in a reservoir fed by the Tigris River "to prevent crops from drying out" during an extended drought in the region. While archaeologists knew of the submerged city, they never had the opportunity to fully survey it. Among the new findings were "a massive fortification with [a] wall and towers, a monumental multi-story storage building and industrial complex." The ancient city dates back to between 1550 and 1350 BCE and is believed to be part of an "extensive urban complex" in the kingdom of Mitanni in Ancient Mesopotamia.[34]

One might think that Turkey, among all the nations of the Mediterranean, is in the enviable position of controlling the headwaters of the mighty Euphrates and Tigris Rivers, and therefore less at risk of running short on water – not so. To get a sense of how grim the heat situation is in Turkey, Baris Karapinar, a professor of climate change policy at the London School of Economics and a lead author of the UN IPCC's fifth assessment report, says that temperatures in Turkey could rise "by seven degrees Celsius over 1950 levels by 2100," vastly overshooting the UN Paris Climate Agreement to hold global temperature to 1.5 degrees Celsius or less. Karapinar says that a temperature increase of this magnitude "would turn much of the Mediterranean region into 'hell,' making parts of it uninhabitable."[35]

The depletion of waters in Turkey isn't just a matter of bad luck. Turkish agricultural policies beginning in the 1980s incentivized cash crops. Although these cash crops have higher yields and propelled the country into the seventh largest agricultural producer in the world and a leading exporter, they require far more water. Unfortunately, growing corn, sugar beet, and cotton, while lucrative on world markets, uses up on average three or four times more rainfall than Turkey receives. The result is that while Turkey boasts a large agricultural sector, "it is devouring 75% of the country's freshwater use," a figure that experts warn is not sustainable. Doğanay Tolunay, a land-use researcher at Istanbul University, points out that these super-thirsty crops have driven up the water use by a third in a decade, a clear example of a failed economic policy.[36]

Faced with a dwindling supply of irrigation water, many farmers are illegally drilling wells to tap a diminishing pool of ground water, which means less water available to replenish lakes, rivers, and wetlands, all of which translates into less surface water for irrigation and human consumption.

Compounding the problem of a precipitous decline of water is the fact that Turkey's ancient irrigation techniques, which rely on open channels and raised canals to deliver water to agricultural fields, are experiencing "losses of 35–60% through evaporation, seepage, and leakage."[37] And then there is the issue of generating hydroelectricity. Yale University's School for the Environment sums up the conundrum:

> Turkey's widespread use of hydroelectric power is sapping water supplies. The world's ninth largest hydroelectric power producer has dammed just about every river in the country, including the iconic Tigris and Euphrates Rivers. While hydropower is a renewable energy source, it dries out aquifers and creates water scarcity downstream of the dam. Reservoirs can lose thousands of liters per second to evaporation.[38]

It wasn't that Turkey was blind to the consequences of damming up every last ripple of the Euphrates and Tigris rivers within its boundaries. In Turkey, as elsewhere in a geopolitical world, it's always a question of environmental assets. The headwaters of the Euphrates and Tigris Rivers are inside the political boundaries of Turkey, and the country has become increasingly occupied with controlling these precious but receding assets to secure its economy and stabilize its governance.

Like Turkey, Iran has embarked on a massive deployment of 541 large and small dams, mostly along the Euphrates and Tigris tributaries, while an additional 340 dams are on the drawing board.[39] The dwindling hope is that the new dams will sequester the remaining waters within Iran and hold off, at least for a time, the inevitable running out of water in the Euphrates and Tigris Rivers.

Hundreds of thousands of people were forcefully displaced in recent decades in both Turkey and Iran with the massive buildout of dams and accompanying hydraulic infrastructure uprooting their way of life. In Turkey alone, many migrated from their rural regions to Turkey's three major cities and especially to Istanbul, where some found new opportunities and still others found only menial work or remained on the public dole for generations.

Today, Turkey hosts the largest refugee population in the world – some four million people, mostly from Syria, but also from other neighboring countries who have fled their homelands because of internal insurgency but also the dwindling supply of water as a warming climate parched their fields.[40] The mass displacement of populations from the rural areas of Turkey as well as the climate and war-driven refugees from Syria and other neighboring countries streaming into Turkey has transformed major cities into behemoths. Istanbul, whose population totaled a mere 967,000 in 1950 stands at sixteen million in 2023 and is one of the largest cities in the world in terms of population within city limits.[41]

All of the hydraulic maneuverings by Turkey have done little to abate its own water deficits. In 2021, Istanbul's reservoir levels dropped down to 20%, leaving its millions of residents with less than 45 days' supply of water. Other regions, at the same time, were equally water stressed by droughts stretching out over two years. Note that half of Istanbul's water supply comes from surrounding cities whose own water reserves run low during droughts, creating a tumbling chain reaction of sorts that puts urban populations at risk all down the line.[42] The unrelenting drought caused near panic, and without any way to alleviate the problem the government "made the director of . . . religious affairs organize 'pray for rain' ceremonies in every city in Turkey."[43]

The issue isn't only about how much water is available in a climate warming atmosphere, but also the quality of that water. A report prepared by the Istanbul Policy Center discusses the knock-on effect of a rewilding hydrological cycle that brings with it floods, droughts, heatwaves, and firestorms. For example, heavy torrential rains adversely affect the quality of freshwater sources by mobilizing:

> pathogens in the environment, carrying them into rivers, lakes, reservoirs, coastal waters and wells [. . . this] changes the flow and content of rivers; it increases the load on sewerage and wastewater treatment plants, causing waste water to mix with drinking water; and it creates water sanitation problems [. . .] heatwaves affect the quality and quantity of water through the thermal and oxygen dynamics and phytoplankton populations in changing water bodies, causing bursts of cyanobacteria.[44]

Istanbul, like other megacities, is hoping against hope that somehow and somewhere down the line a warming climate will reverse itself to the pre-global warming climate, which of course will not happen. The city would be smart to look at Cape Town, South Africa, with its four million inhabitants, which sleep-walked its way to the brink in 2018 by ignoring three consecutive years of drought that had depleted its reserves. The situation became so dire in Cape Town that city officials began publicly announcing what they called "Day Zero," the date when there would be no more fresh water flowing through the pipes, and all the faucets would shut down, leaving its millions of citizens stranded and helpless.

While officials didn't call for a city-wide prayer session as Turkey did in hope of divine intervention, a number of bizarre proposals were introduced, and at the top of the list was the idea of towing icebergs from the Antarctic to Cape Town to provide water for the population. The proposal was first floated during an earlier drought between 2015 and 2018 and took on a second life in the run up to Day Zero. The proposal died, however, when a U.S. scientist ran the numbers and concluded that during the 2,500-kilometer journey from the Antarctic Ocean to Cape Town, the 300-meter iceberg would melt, leaving only one percent of its initial volume, enough water to last for four minutes for thirsty Cape Town inhabitants.[45]

Renewables to the Rescue and a Second Life for the Mediterranean Eco-Region

Energy lies at the center of the water–energy–food nexus, and it's here that Turkey finds itself too trapped by its reliance on fossil fuels to move the economy forward. Fossil fuel imports make up approximately 70% of Turkey's energy consumption.[46] Other countries of the Middle East – Saudi Arabia, Iran, Iraq, Kuwait, United Arab Emirates, Egypt, Qatar, Oman, and Yemen – where oil revenues are a mainstay of their economies and export markets, face a similar entrapment. Although there's talk about lessening reliance on fossil fuels, precise targets have yet to be set.

The recent discovery in Turkey of the giant Sakarya gas field in the Black Sea and other efforts to expand gas and oil exploration and production paints a picture of digging in rather than digging out of fossil fuels as the country's primary energy.[47] Despite occasional lip service to reduce its dependence on coal, output has only "been falling slightly over the past ten years."[48] If Turkey and other Mediterranean countries continue to rely on fossil fuels as a dominant source of power for both domestic production and imports and exports, they will further box themselves into short-term reliance on oil, coal, and natural gas at the expense of the further warming of the climate and to the detriment of their economies.

What makes the situation particularly troubling from a strictly economic perspective is that in 2019 the levelized cost of both solar and wind energy dipped below the cost of nuclear, oil, coal, and natural gas, and the fixed cost continues to drop on a plummeting exponential curve, while the marginal cost of generating the power is near zero. The sun and the wind have never sent a bill. Once the technology is installed, the only cost is maintenance and replacement. Clearly the market is here, which is why solar and wind together account for most of the new expansion of global power generation in the world. The International Energy Agency projects that solar and wind "are set to account for over 90% of global electricity expansion over the next five years, overtaking coal to become the largest source of global electricity by 2025."[49]

The powerhouse countries of the European Union that abut the Mediterranean understand the game-changing market potential of solar and wind and are leaping into the green energy future. In 2023 Spain announced that it had set a new target of reducing greenhouse gas emissions by 32% by 2030, from a previous goal of 23%, because of the plummeting cost of solar and wind in the market. The country is set to install an eye-popping 56 gigawatts of solar power and 32 gigawatts of wind power in just the next seven years.[50] Spain is also rushing forward with green hydrogen, forecasting eleven gigawatts of electrolyser capacity by 2030, sharply up from its earlier projection of only four gigawatts.[51]

Even smaller EU countries in the Mediterranean are quickly ramping up green energy. In October of 2022, Greece announced that, "For the first time in the history of the Greek electricity system" 100% of the power generated was from renewable energy for five hours during the day. Greece is targeting 70% of its electricity from renewable energy by 2030.[52] These first-mover EU countries of the Mediterranean enjoy close ties with other countries that share the Mediterranean Sea. European energy companies' expertise and capital are being ramped up in the Mediterranean, and even in the oil rich countries of the Middle East, and are beginning to lure the stalwart oil meccas into renewable energy states, if only marginally.

Already, more than eleven trillion dollars have flooded out of the global fossil fuel complex, as investors have come to realize that what lies ahead is trillions of dollars of "stranded assets."[53] The International Energy Agency (IEA) projects that global demand for coal "falls back within the next few years, natural gas demand reaches a plateau by the end of the decade, . . . [while] oil demand levels off by the mid-2030s . . ." The IEA notes that "global fossil fuel use has grown alongside GDP since the start of the Industrial Revolution in the eighteenth century: putting the rise into reverse will be a pivotal moment in energy history."[54] Well-said!

Thus, the Mediterranean region, due to its unique geographic situation and its massive oil and gas reserves finds itself betwixt and between a dying fossil fuel culture – once a blessing and asset, but now a curse and liability – and a promising renewable energy society now scaling. Turkey, again, is a good case study on the conundrum facing so many Mediterranean countries that are awash with fossil fuels. Turkey is an intermediary in moving oil and natural gas between Asia and Europe. William Hale, a professor of politics and international studies at the School of Oriental and African Studies at the University of London, makes the point that "Turkey occupies an important geo-strategic position between the world's major exporters of oil and natural gas in the Middle East, the Caspian Basin, and Russia, on the one hand, and Europe on the other. As a result, five important international pipelines currently cross Turkey's territories."[55]

The Kirkuk–Ceyhan pipeline brings oil from Iraq's oil fields to Turkey's Mediterranean Coast for both internal consumption and export elsewhere. The Turkey–Iran pipeline carries gas from Tabriz to Ankara, where it connects with the domestic Turkish gas network. The Southern Gas Corridor connects Azerbaijan's gas fields with Turkey, Georgia, Greece, and Italy. The Blue Stream gas pipeline runs from Izobilny in Russia under the Black Sea and onto Ankara. Finally, the TurkStream gas pipeline carries Russian gas under the Black Sea through Russia and Turkish maritime zones into Turkey and over into Greece, with extensions into Bulgaria, Hungary, Slovakia, Austria, and Serbia.[56]

In the short run, it's easier to stay enmeshed in a dying fossil fuel matrix, even knowing that this entire fossil fuel pipeline infrastructure is becoming a stranded asset as early as the late 2020s. The continued reliance on fossil fuels is literally adding fuel to the fire with further emission of global warming gases heating up the climate across the Mediterranean, imperiling the whole of the Mediterranean eco-region and putting in jeopardy its millions of inhabitants.[57]

On the other side of the impending cataclysm lies an opportunity to rethink the future of a shared Mediterranean eco-region by engaging in a collaborative arrangement to transition from a dying fossil fuel-based hydraulic civilization to a zero-emission resilient renewable energy society. For the hundreds of millions of human beings who call the Mediterranean their home, this invitation is likely too good to pass up, but might not happen.

Once again, let's look at Turkey and what the International Energy

Agency has to say about the country's future prospects. Perhaps the most significant finding in the IEA 2021 report on Turkey's energy portfolio is that hydropower in the country will reach its capacity by 2023 and that "additional hydro will play a limited role after 2023." As the climate models predict, the Euphrates and Tigris Rivers will likely dry up in the next twenty years or so, leaving a vacuum that will have to be filled by either fossil fuels or nuclear power plants, both of which require massive withdrawals of water to generate electricity and cool thermoelectric power plants, or by reliance on the superabundance of onshore and offshore solar and wind energies buttressed by tidal, wave, and geothermal energies. It's a "no brainer."[58] But the issue is not if, but when – waiting too long by remaining embedded in an archaic and dying fossil fuel and nuclear power infrastructure might well run out the clock, making it too late for the transition to the new energies and accompanying infrastructure of a budding Age of Resilience.

Here are the numbers – and they are encouraging – potentially placing Turkey as a lead mover in the Mediterranean in making the transition in the region to a resilient Third Industrial Revolution post carbon infrastructure. While Turkey currently spends $50 billion annually in imported oil, gas, and coal, placing a straitjacket on an already weakened economy plagued by runaway inflation, the country is also sitting on a solar goldmine, and is the "second best country in Europe for solar power generation based on solar intensity and availability."[59]

As of 2023, the total installed solar energy capacity had delivered far beyond expectations, signaling a possible seismic shift into a new energy era. With an average of 7.5 hours of sunshine a day, rooftop solar power plant potential alone is estimated to exceed an incredible twenty gigawatts by the end of the decade, positioning Turkey as a principal player in solar energy power generation in the Mediterranean. Solar is already scaling in 78 provinces across Turkey, offering the prospect of provisioning distributed solar power generation across every region.[60] What's triggered this sudden rush to install solar power is inflation, now at a record high in Turkey, brought on by the collapse in the currency and the spike in global energy costs resulting from the Russian invasion of Ukraine. Homeowners and businesses that can afford it are quickly installing solar panels to get around high electricity costs and inflated global fossil fuel energy prices, sparking a meteoric rise in solar energy companies and the solar energy market.

In one two-week period alone in 2022, over 300 companies applied for approval to install solar panels, according to the state-run Anadolu Agency. The head of Turkey's regulation agency reported that "companies made $110 billion-worth of investment applications for renewable energy production."[61] The Turkish government sweetened the pot by easing the process that enables homeowners and companies to sell their surplus power back to the grid.

With money tight, banks have jumped into the fray, offering affordable green financing. The European Bank of Reconstruction and Development has also stepped up, setting aside $522 million in 2022 to finance green investment in Turkey.[62] The bull market in solar has encouraged companies to invest in the manufacturing and installation of solar panels and utility scale solar units across the country. To get a fix on the scale of the transition thus far, "the installed solar energy capacity of 8.3 gigawatts (GW) in Turkey is calculated to exceed 30 gigawatts by 2030 with roof and field type projects."[63] The installation of just one gigawatt of solar-generated electricity is enough to power 750,000 homes.[64]

While Turkey is blessed with an abundance of potential solar power unrivaled in the Mediterranean, its offshore wind opportunity is equally impressive. The Global Wind Energy Council World Report of 2022 takes note that Turkey, along with Azerbaijan, Australia, and Sri Lanka, boasts the highest offshore wind potential in the world.[65] Much of Turkey's wind potential lies in the Aegean region of the Mediterranean Sea. Although installing offshore wind is more expensive than onshore wind, the cost is balanced by the higher energy production. Wind energy roadmaps are currently being prepared with the goal of producing 20 gigawatts of installed wind by 2030 in addition to 10 gigawatts already installed. The Izmir Development Agency estimates Turkey's total offshore wind potential at 70 GW.[66]

Offshore wind could even reach an impressive 75 gigawatts and, when combined with installed solar electricity potential of 30 gigawatts and onshore wind electricity potential of 48 gigawatts,[67] Turkey is set to make a transition from a fossil fuel-based economy to a renewable energy economy well before midcentury.[68] European and other international companies have rushed to set up operations in Turkey, some going it alone and others partnering with homegrown solar and wind companies already on the ground. By 2021, more than 3,500 companies employing 25,000 workers were operating in Turkey installing on land wind turbines and exporting parts and entire wind turbine systems abroad. The American company, TPI Composites, for example, which produces high-quality composites that go into wind turbines, has two plants operating in the Izmir region with a workforce of approximately 4,200 employees. The company serves the Turkish market and exports to Europe and the Middle East.[69]

New reports are suggesting the possibility of combining solar and wind on offshore platforms in the Aegean region of the Mediterranean. These studies point out that combining solar and wind on floating and anchored offshore platforms comes with a number of advantages. For starters, solar irradiance offshore is higher than onshore. The thought is that while combining solar, wind, and wave energy means higher initial costs, the gains are made on the backend by generating more renewable energy and in a more balanced distributed way saving on costs, while at the same time ensuring a dependable supply of renewable energy.[70]

There is one more feature to the joint deployment of offshore wind and solar generation platforms that is speeding the transition away from a fossil fuel-based infrastructure and to a resilient zero emission green energy infrastructure. Hydrogen is beginning to come into play, particularly in the Mediterranean. Hydrogen is the most ubiquitous element in the universe. Here on Earth, hydrogen has long been utilized in industry mostly as a catalyst in ammonia production, oil refining, and methanol production, all critical to a number of industrial processes.

In recent years, there has been growing interest in extracting hydrogen from coal and natural gas and using it to power industrial processes and as a fuel for transport, among other things. These uses are called grey and blue hydrogen. Hydrogen can also be distilled from water. In this instance, solar and wind-generated electricity are used to separate hydrogen and oxygen molecules in water and use the stored hydrogen as a liquid or gas for use in industries and/or as a fuel for transport. It's called green hydrogen.[71]

While the renewable energy industry is ramping up green hydrogen as a way to store renewable energy and transport it via pipelines for road, rail, and water transport and for use in heavy industries, the problem is that the coal and natural gas industries are also supporting hydrogen but extracted from fossil fuels and categorizing it as a transition fuel on the way to green hydrogen down the line – thus delaying the transition into a zero-carbon emission economy.[72]

In September 2022, Aquaterra Energy and Seawind Ocean Technology entered into a partnership to develop the world's largest offshore floating wind and green hydrogen production project off Italy in the Mediterranean Sea. The venture will have the capacity of 3.2 gigawatts and will begin operations in 2027. The green hydrogen produced offshore will be transported by pipelines to shore and by vessels, and from there to global markets.[73]

With all of the buzz around offshore wind, solar, tidal, and green hydrogen across the Mediterranean Sea to transition the nations of the region into the Age of Resilience, we need to keep in mind that the Mediterranean is an eco-region and whatever use is made of floating and/or anchored offshore wind/solar/hydrogen platforms will need to be assessed for their environmental impact from the onset.

One of the anchors of the European Union is the Precautionary Principle, which states that "where scientific data do not permit a complete evaluation of the risk, recourse to this principle may, for example, be used to stop distribution or order withdrawal from the market of products likely to be hazardous." The principle has wide application across every field, including human health, animal and plant health, and the environment at large.[74] The Precautionary Principle can be invoked if three preliminary conditions are met: (1) The identification of potentially adverse effects. (2) Evaluation of the scientific data available. (3) The extent of scientific uncertainty.[75]

Interreg Mediterranean is a project financed by the European Union Regional Development Fund. The organization has published a detailed guide to assessing the environmental impacts of marine renewable energy potentials in the Mediterranean region. The Interreg Mediterranean realizes that marine renewable energy is essential to transitioning the Mediterranean to a post carbon era and a flourishing of marine life in the region but stresses that the deployment of wind, solar, and green hydrogen renewable energies "must be achieved without posing any additional threats to biodiversity and ecosystem conservation."[76]

While offshore solar platforms pose few environmental risks, wind turbines raise a specter of concerns, including bird collisions and continuous noise disturbance, injuring marine mammals as far as hundreds of kilometers away. The laying of cable on the sea floor, while less problematic, may also negatively affect the overall ecological footprint. Erecting stationary as well as floating wind platforms may upset ecosystem dynamics. The Interreg Mediterranean Directive argues that the most effective method for limiting negative impacts is spatial segregation, i.e. "careful initial site selection to avoid high conservation valued areas: this would exclude [marine protected areas] as potential locations for [offshore wind farms]."[77]

A significant consideration in operating offshore windfarms according to Interreg Mediterranean is to "support the development of new technologies for floating wind turbines (floaters, anchors) capable of operating in deep waters and/or far away from shores and design them in order to minimize their impacts on marine ecosystems and biodiversity."[78]

Since weaning the Mediterranean off a fossil fuel infrastructure is critical to eliminating global warming emissions, while protecting marine life in the Mediterranean Sea is of equal importance, finding the most effective way to ensure that both goals are met is a balancing act of great importance. By keeping the Precautionary Principle front and center in all future development of offshore wind, solar, and green hydrogen power generation, and applying ecological modeling accompanied by ever-more sensitive and benign offshore methods to harnessing wind, solar, waves, and tides, will ensure a thoughtful restoration of the Mediterranean eco-region.

Solar and wind energies are upending the 200-year reign of fossil fuel-based industrial energy, especially in the EU and China and beginning in the U.S.; less so in other parts of the world. The International Renewable Energy Agency says that, by 2050, 90% of electricity can come from renewable energy.[79]

While the Mediterranean Middle East has the largest reserves of fossil fuels and drives the world market, these countries too are beginning to make the turn to green renewable energies. But, even with the turn to renewables across the Mediterranean, our species will have to confront an even bigger problem – the diminishing amount of fresh water available for both human consumption and our fellow creatures and for food production. To put this in perspective, Day Zero is approaching in ecosystems on every continent. Consider this statistic. The World Bank reports that "in the last 50 years, fresh water per capita has fallen by half," partially because world population has more than doubled from 3.7 billion in 1970 to 7.8 billion in 2021. But it's also because in many regions of the world, global warming has led to chronic drought, drying up lakes, rivers, and underground aquifers everywhere, but especially in arid and semi-arid regions. The fact that the amount of water available per capita has fallen by half in fifty years is perhaps the most devastating statistic of all and an indication of how close our species is to a swift and inevitable decline in human numbers over the course of the twenty-first century.[80]

Water, Water, Everywhere, But Not Any Drop to Drink . . . Until Now

"Forty percent of the world's population live within 100 kilometers of the coast."[81] In the *Rime of the Ancient Mariner*, a poem by Samuel Taylor Coleridge, a sailor on a becalmed ship, stalled in the open seas, laments, "water, water, everywhere, but not any drop to drink." A reprieve of sorts is emerging in the form of new desalination technologies that are scaling in the Mediterranean with its hundreds of millions of people running precariously short of available fresh water. Water desalination involves extracting salt from saline water to provide fresh water. The older process is thermal desalination which uses fossil fuel energy to change "saline water into vapor . . . when condensed, this vapor forms a high-purity distilled water."[82] The second and relatively new process for the desalination of water is reverse osmosis, which separates fresh water by straining seawater through a membrane at high pressure.[83]

Most of the 17,000 desalination plants rely on the thermal process using fossil fuels to heat the salt water and create the vapor or steam that is free of the salt, minerals, and other contaminants. The Middle East accounts for nearly 50% of the desalinated water produced in the world. Saudi Arabia, the United Arab Emirates, Israel, Kuwait and Qatar together make up approximately one-third of the world's desalination plants.[84] But with solar and wind energy now cheaper than all of the fossil fuels, the push is on to switch from fossil fuels to renewable energy-powered desalination.[85]

By 2040, about 75% of the fresh water produced by desalination in the Mediterranean Middle East will be from membrane-based osmosis. But as long as gas and oil remain subsidized by the oil-producing states in the region, it's likely that much of the process will continue to rely on fossil fuel-generated electricity, further contributing to the warming of the climate in the Mediterranean.[86]

Today, renewable energy deployed in desalination is a skimpy one percent. That's likely to quickly change even in the oil-producing countries in the Middle East now that solar and wind energy are cheaper than fossil fuels.[87] The Global Clean Water Desalination Alliance is made up of energy and desalination industries, water utilities, governments, financial institutions, and R&D and academic institutions who have banded together "to scale up the use of clean desalination technologies."[88]

Many in the upper echelon of the water and energy sectors are bullish in the belief that solar and wind-generated electricity used in the desalination process will more quickly scale and leave fossil fuel electricity behind, noting that "the cost of solar-powered thermal desalination is expected to drop by 40% or more by 2025."[89] Carlos Cosin, CEO of Almar Water Solutions, part of Abdul Latif Jameel Energy, like others looking at the fast downward cost of solar and battery energy technologies says, "for me it's clear: the future of desalination will be with renewables. It's only a matter of time in the Middle East region."[90]

In December of 2022, Egypt announced a massive $3 billion plan to deploy 21 water desalination plants using renewable energies. With the Nile River quickly drying up, putting its roughly 105 million citizens at extreme risk of edging ever closer to Day Zero, the plan will eventually generate 8.8 million cubic meters of desalinated water daily, at a cost of $8 billion, providing a new water venue for its population.[91]

The Spanish global energy and water company, ACCIONA, has teamed up with the Kingdom of Saudi Arabia to deploy four desalination plants and three wastewater plants in addition to the two plants already online. Altogether, the plants will provide "enough [water] to supply 8.3 million people, almost one-quarter of the country's population."[92] On the other side of the world, Western Australia has mandated that all new future desalination plants be powered by renewable energy.[93]

Recent breakthroughs in renewable desalination offer the prospect of greatly reducing the cost and delivery time of desalinated water, opening up the possibility of providing fertigation on agricultural lands facing severe droughts. In 2021, researchers at the University of Texas and Penn State University reported that they were able to upgrade the efficiency of membranes at the nanoscale by 30–40%, allowing them to clean more water with far less energy – the kind of breakthrough that's necessary if countries are to produce a substantial volume of fresh water to assist in fertigating and irrigating agricultural fields.[94]

Already, three hundred million people in 150 countries use desalinated water from some 17,000 desalination plants for their drinking water, bathing, cooking, washing clothes, and other household and business practices. With the technology improving and the cost dropping on an "exponential curve," it's not impossible to imagine these figures tripling, quadrupling, and more in coming decades, with a sizeable portion of the human race relying on desalinated ocean water. The big question is exactly how fast the cost of desalinated water will drop in the coming decades, allowing for home and industry use and the irrigation of agricultural land on every continent now facing serious droughts. A decade ago, that would have been sheer fantasy. Not now.

The fly in the ointment is that even with the ever-cheaper cost of using solar and wind energy in the reverse osmosis process there is the sticky and not inconsequential problem of what to do with the concentrated brine left behind, which is generally dumped back into the sea, posing the risk of environmental damage to marine ecosystems. Fortunately, chemists and engineers are testing a myriad of approaches to turn the problem into an opportunity by experimenting with a variety of ways to extract the valuable metals from the brine for important industrial uses, while protecting the sea water and reducing the bottom-line costs of using the elements and metals for commercial purposes.

The Massachusetts Institute of Technology (MIT) is among the many institutions experimenting with various options. In a paper published in the journal *Nature Catalysis*, MIT research scientists noted that the desalination industry uses a lot of sodium hydroxide – caustic soda – to pretreat seawater being readied at the desalination plants, to lessen the acidity of the water and prevent fouling of the membranes used to filter out the salty water, which often causes serious interruptions in the osmosis process.[95] Were the extraction of the sodium hydroxide from the brine to be done onsite, instead of purchased in the open marketplace, the savings would be significant. And, because the amount of caustic soda extracted onsite would exceed the requirements used in the osmosis process, the surplus could be sold in the market for use in the chemical industry, while reducing the amount of brine discharged back into the sea.

Chemistry laboratories across the world are also turning their attention to the extraction of valuable elements from hyper-saline brines used in the desalination osmosis process – including "molybdenum, magnesium, scandium, vanadium, gallium, boron, indium, lithium, and rubidium." Most of these elements, which up to now have been mined from medium to high grade ores that are becoming increasingly depleted, are making their extraction from sea brine more attractive.[96]

What is clear is that the desalination process is on a fast-moving exponential curve. Recall that in the early years of pre-market tinkering of computer prototypes back in the 1940s, then president of IBM Thomas Watson remarked, "I think there is a world market for maybe five computers."[97] What Watson didn't anticipate is exponential curves. By the late 1950s, Gordon Earle Moore, the cofounder of Intel, ran the numbers and was taken aback to find that Intel engineers were successfully doubling the number of components on Intel chips every 24 months, allowing for more powerful computing, all the while making their computers ever cheaper – to wit, the cost of computer chips were on an exponential curve and continue to be so.[98] By 2023, there were 6.7 billion smartphones in use, each of which had more computing power than sent our astronauts to the moon.[99]

As already mentioned, solar renewable energy has been on a similar exponential falling cost curve for nearly a half century. Wind is also experiencing a plunging exponential cost curve, making the levelized cost of these renewable energies cheaper than fossil fuels and nuclear power. A comparable exponential curve is already taking shape in desalination processes.

Many farm regions are beginning to turn to desalinated water to fertigate crops as their rivers and aquifers dry up. In Spain, for example, farmers have used 22% of the country's desalinated water for fertigation. Kuwait has used 13%, Italy has used 1.5%, and the U.S. has used 1.3% of the desalinated water for fertigation.[100]

Desalination of ocean waters, even at scale, is not a quick fix that's going to allow our species to return to the old normal. Those days are forever gone. What desalination will do, along with the many other changes we will need to make in the way we live, is ease the transition as our species makes its way to a new more spartan but robust way of life where a hedonist utilitarianism gives way to a flourishing repatriation with our animated planet.

CHAPTER 6

Location, Location, Location
The Eurasian Pangaea

There is a tried-and-true aphorism familiar to every real-estate developer ... that one's good fortune – or lack thereof – all depends on location, location, location. And nowhere is that more so than in what's likely to be one of the most valuable locations in the world today – and, as in ancient times, it's the Mediterranean region. How might this be so in a region whose once majestic rivers – the Euphrates, the Tigris, and the Nile – spirited civilization into existence but now are quickly drying up? As in the past, it's about location, location, location, but not tied to its once great rivers but to its strategic position between the continents of Europe, Africa, and Asia.

The Bridge to Europe and Asia

It's worth noting that "in 2022, China was the third largest partner for EU exports of goods (9.0%) and the largest partner for EU imports of goods (20.8%)."[1] The long and short of it is that Eurasia is, for all intents and purposes, a single extended landmass and potential seamless market, and Turkey is "literally and figuratively" the connecting bridge between both the EU and China and everything in between (the four bridges in Turkey that connect Europe and Asia are the Bosphorus Bridge, the Fatih Sultan Mehmet Bridge, the Yavuz Sultan Selim Bridge, and the 1915 Canakkale Bridge).

The first move to establish a dynamic new twenty-first-century seamless Eurasian market place came in 2013 when President Xi Jinping of China floated the idea of a contemporary version of the Ancient Silk Road on a visit to Kazakhstan. He called the proposal the Belt and Road Initiative (BRI) and suggested that the countries of Asia and Europe be conjoined for the purposes of trade, by establishing new land and sea routes across and around the supercontinent and extending to Africa, whose five North African countries – Morocco, Algeria, Tunisia, Libya, and Egypt – also abut the Mediterranean Sea.

The Belt and Road Initiative, sometimes called the New Silk Road, has already begun deploying projects across Eurasia on the way to becoming among the biggest infrastructure deployment and trade initiatives in world history.[2] As of early 2022, 146 nations along with 32 international organizations have signed cooperation agreements for the BRI.[3] Notably absent are the United States, Canada, Mexico, and most of the countries of Western Europe. The Belt and Road Initiative is currently geared toward investing in a Eurasian infrastructure but eventually looking toward a global deployment of a twenty-first-century infrastructure to include roads, bridges, power lines, transport corridors, airports, and railroad lines across the world.

The World Bank estimates that the BRI is likely to increase global trade in participating countries by 4.1%, while cutting the cost of trade by up to 2.2%.[4] The London-based Centre for Economics and Business Research says, if fully deployable, the BRI interconnect is likely to add 7.1 trillion dollars in GDP per annum by as early as 2040 while "reducing both transport and other frictions that hold back world trade."[5]

The massive interconnective infrastructure will require expenditures of at least 900 billion dollars per annum over the next decade – or 50% above current infrastructure spending.[6] As of 2017, the initial infrastructure rollout was expected to cover 68 countries and 40% of global GDP.[7] There will be both a land and marine route across Eurasia. The land route will be composed of seamless road and rail interconnects for passengers and cargo. The maritime silk road ocean route will run from the China coast across the waters of Southeast Asia, South Asia, and Africa, ending up in the Mediterranean.

Turkey, fully cognizant of its strategic advantage as the lynchpin bringing Asia and Europe together in a single continental matrix for the purpose of commerce and trade, has announced its own initiative called the Trans-Caspian East–West–Middle Corridor, also known as the Middle Corridor, which starts in China and passes through Central Asia and the countries of Kazakhstan, Kyrgyzstan, Uzbekistan, and Turkmenistan and across the Caspian Sea, and from there heads across Azerbaijan and Georgia to Turkey and the Mediterranean. Turkey has already brought together Azerbaijan and the Central Asian republics in a cooperative protocol, and other Mediterranean countries are beginning to sense the opportunity.[8]

Whether the nations that surround the Mediterranean can close the deal and ensure that their ports will serve as the logistics hubs connecting the Eurasian supercontinent is the great political and economic question of the century, and an historic opportunity, but still very much in doubt. Still, the Mediterranean Region has seen a 477% increase in cargo processed by its ports – a strong number by any calculation. The Mediterranean Sea currently harbors 20% of all global shipping, and there is a growing sense that "there is no alternative route as efficient as the Mediterranean Sea to connect Asia to Europe."[9]

Unfortunately, Mediterranean ports have been sluggish to modernize and take advantage of the unique opportunities now surfacing. Of late, however, eyes have begun to focus on Eurasian logistics and supply chains meeting in the ports of the Mediterranean Sea and sealing the deal to connect Europe and Asia in a shared economic zone. What has changed the equation and awakened the sleepy ports of the Mediterranean Sea is the increased market share of what's called transshipments – "the shipment of goods to an intermediate destination, then to another destination."[10]

A recent study by the German Marshall Fund has drawn attention to this previously ignored category of transshipments and the role that the Mediterranean ports will likely play in the moving of cargo between Europe and Asia. The report notes that "the spike in transshipment activities has raised the strategic importance of the Mediterranean Sea (i.e. the intermediate destination) as a connecting market but tends to serve the global more than domestic market, and particularly the world's largest exporters than the countries where transshipment hubs are located." The report goes on to say that the meteoric rise of transshipments "explains why the Mediterranean is so importantly placed within the Chinese Belt and Road Initiative (BRI)."[11]

The European Union, which has been slow to enter the arena, announced its flagship Global Gateway initiative in 2021 designed to compete with the BRI. Some 70 projects have already been approved with an initial commitment of 300 billion euros in public and private funds to be allocated by 2027 in deploying twenty-first-century infrastructure projects including a hydrogen project in Kazakhstan, a transport link in Central Asia, and a hydropower plant in Tajikistan, along with projects in Mongolia. Yet, there has been little public discussion at the European Commission-level around infrastructure across the Mediterranean ecosystem that will connect Europe and Asia as a seamless economic space – that is likely to come more quickly with the easing of the COVID-19 pandemic.[12]

In one respect, the EU is better positioned, at least geopolitically, than one might suspect to engage both competitively and cooperatively with China on facilitating the deployment of a Eurasian infrastructure for collaborative trade. The ace in the hole may be the Union for the Mediterranean, which was founded in 2008 with the objective of encouraging greater institutional cooperation between the 27 EU member states – of which Spain, France, Italy, Greece, Croatia, Slovenia, Cyprus and Malta abut the Mediterranean and fifteen other member states who share the ecosystem – Albania, Algeria, Bosnia and Herzegovina, Egypt, Israel, Jordan, Lebanon, Mauritania, Monaco, Montenegro, Morocco, Palestine, Syria, Tunisia, and Turkey.

Joint cooperation between the European Union and the Union for the Mediterranean (UfM) will need to include environmental governance, the improvement of ports and railways connecting the regions, the push for renewable energies, the promotion of higher education and research, and a commitment to increasing small and medium-sized enterprises. In 2011 the Union for the Mediterranean announced its first project – the deployment of a seawater desalination plant in Gaza. And, in 2016, the UFM turned its attention to formulating a collective Mediterranean climate agenda to steward the Mediterranean region.

The bringing together of the EU member states and the member states of the UFM in a shared institutional venture – now with fifteen years of joint political engagement in the field – gives the region a strong position to establish both a competitive and cooperative relationship with China's Belt and Road Initiative and the EU's Global Gateway. While healthy competition will spur the deployment of infrastructure across Eurasia, all the parties will also need to cooperate and wrestle with the establishment of formal codes, regulations, and standards of conduct to address security issues and engage in collaborative efforts to mitigate climate change, with the goal of transforming the

Eurasian supercontinent into a zero-emission landmass and a shared super commons.

In the long run, it will be important to keep in mind, as already mentioned, that the EU and China and all of the countries in between do indeed share a common landmass. And, if geography is destiny, both will likely be each other's principal trading partner far into the future. The future well-being of Eurasia will ultimately depend on the joint efforts of the two continents to address climate change and work collectively to steward their shared landmass.

The inklings of a seamless Eurasian infrastructure are coming into view, but to date there remains a misunderstanding of the nature of infrastructure paradigms – of both their ontology and instrumentality. All three parties – the European Union, the People's Republic of China, and the Union for the Mediterranean are consumed with the idea of pursuing a labyrinth of standalone projects of strategic importance – often helter-skelter – with the hope that they will connect down the line and create a streamlined infrastructure that will enable seamless commerce and trade across the supercontinent. Each views the infrastructure rollout from a geopolitical lens – hoping to exact a competitive advantage – when in fact the infrastructure that is emerging requires a biosphere lens.

What's missing from all of the calculations is that the continued reliance on geopolitics and the securing of market advantage is inimical to the next great transformation in history, whose modus operandi is a collective biosphere politics – that is, the shared stewardship of the global hydrosphere and accompanying lithosphere, atmosphere, and biosphere – the planetary umbrella under which we all survive and flourish. That doesn't mean that robust competition is without merit, but that it has to be within the larger context of a cooperative responsibility to steward the Mediterranean as an ecological commons.

The Mediterranean constitutes a distinct bioregion and, from a strictly ecological perspective, is recognizable as a natural commons. The European Environment Agency defines a bioregion as, "a territory defined by a combination of biological, social, and geographic criteria, rather than geopolitical considerations; generally, a system of related, interconnected ecosystems."[13]

Within the context of biosphere politics, governance begins at the bioregional level and the protection of local ecosystems and radiates out in various concentric circles, enveloping the entirety of the Eurasian landmass. This rethinking of the "Belt and Road," "Middle Corridor," and "Global Gateway" initiatives from the perspective of biosphere politics rather than geopolitics is already occurring, albeit piecemeal and tentatively at best across the nations that share the Mediterranean.

In 2004, the Spanish autonomous region of Catalonia, the French region of Occitania, and the Spanish government of the Balearic Islands established the Euroregion Pyrenees Mediterranean, the first official bioregion in the Mediterranean. The region has fifteen million inhabitants with a population greater than twenty of the 27 nations of the EU and covers a territory of 110,000 square kilometers – making the bioregional governing space larger than seventeen of the EU nations. The combined GDP of the bioregion as of 2019 was EUR 440 billion, equivalent to Austria's GDP.[14]

This first of a kind bioregional governance in the Mediterranean has set forth an ambitious agenda that it hopes will spur bioregional governances across the Mediterranean and from there around the world. Its mission statement is bold and a crucial first step in partially shifting governance from a strictly geopolitics orientation wedded to protecting nation states' sovereignty to include a new "biosphere politics" that crosses national boundaries, allowing regions to share responsibility of stewarding their common ecosystem. This new biosphere politics governing entity leans forward with a powerful green line strategy that follows a transformational new political agenda. The Euroregion Pyrenees Mediterranean governance provides a generally accepted description of the nature of bioregional governance that fits a wide range of initiatives:

> The Euroregional social, economic, and territorial developmental initiatives promoted will be conditional on the conservation of natural ecosystems; i.e. the protection of biodiversity, water, forests, and marine ecosystems. We want to address the challenges arising from climate change by acting from a bioregional perspective going beyond a working approach that is centered on administrative boundaries. We will work to provide innovative solutions on the ground, addressing the most necessary transformations, with the mobilisation and involvement of local actors.[15]

Expediting the shift to bioregional governance in the Mediterranean is going to require leveraging government funding as well as investment from the private sector, particularly from pension funds, insurance funds and banks, and private equity. Interreg Euro-Med, a European territorial corporation program co-founded by the European Union will provide funds to public administrations, investors, and private and civil society organizations to address climate change. All of the Interreg Euro-Med policies, initiatives, programs, and projects are related directly to addressing a warming climate and revitalizing the Mediterranean ecosystem by encouraging cooperative and commons governance between adjoining regions. Sounds nice – but when theory meets practice a more sober reality kicks in.

Although the objectives, targets and goals are lofty, the funds allocated are far too little to address the magnitude of the challenge. The fact is that the whole of the Mediterranean infrastructure – its hydropower dams, power lines, mobility and logistics corridors, residential, commercial, and industrial building stock, and conventional farming practices – are stranded assets. This Holocene infrastructure was not designed to operate in a dramatically warming planet plagued by powerful atmospheric rivers in the winter, unleashing massive snowfalls, apocalyptic floods in the spring, extended droughts, life-threatening heatwaves and devastating wildfires in the summer and destructive fall hurricanes and typhoons.

A crisis of this magnitude will require a more intimate engagement, mobilizing entire communities across shared bioregions. And here's the rub. The climate crisis is forcing a struggle between two dramatically opposed political ideologies – a standoff between an entrenched geopolitics and an emerging biosphere politics. Nation states wedded to the conventional geopolitical model are closing ranks behind sovereign borders and engaging in increasingly open conflicts in a last-gasp attempt to hold on to a dying fossil fuel urban hydraulic civilization. Yet, at the local level regions that abut a common ecosystem are increasingly engaging in cross-border biosphere politics, realizing that their future depends on stewarding and revitalizing their shared natural milieu.

The Euroregion Pyrenees Mediterranean bioregional governance and others still in the gestational stage have much to learn. In this regard all of these budding bioregional governance initiatives in the Mediterranean would benefit by examining an earlier effort, still morphing between Israel, the State of Palestine, and Jordan, to establish a bioregional governance across their adjoining sovereign territories and common ecosystem. The cross-border effort initiated by EcoPeace Middle East, an NGO founded in 1994 in Israel, would have all three governments share a water–energy nexus to the benefit of each in a bioregional governance. The plan amounts to a "Middle East Green–Blue deal" and it would have Israel provide desalinated water at scale to Palestine and Jordan, and Jordan to provide utility-scale solar to both Israel and Palestine – what its advocates call a win-win shared bioregional governance.

All three parties have much to gain and little to lose in this sweetheart bioregional governance of an ecosystem. Israel is far ahead of its Mediterranean neighbors in scaling sea water desalination. By 2022, more than 85% of its drinkable water was produced by desalination, giving Israel an edge over other Mediterranean countries in converting salt water to fresh water at scale and at a low market price. Jordan, facing increasing water scarcity, has abundant solar potential with an average of 300 days of sunshine a year.[16] Jordan's landmass is also 4.1 times the size of Israel and can export abundant utility-scale solar energy cheaply to both Israel and Palestine, providing near zero-emission electricity to power their economies and society. This is the way EcoPeace sizes up the plan:

> The water–energy nexus (WEN) is EcoPeace's flagship project for climate change adaptation and mitigation, designed to create a regional desalinated water-solar energy community among Jordan, Israel and Palestine that would result in healthy and sustainable regional interdependencies. Israel and Palestine would produce desalinated water, and sell it to Jordan, while Jordan sells Israel and Palestine renewable energy, thereby enabling each partner to harness comparative advantage in the production of renewable energy and water.[17]

On a very practical level, the bioregional governance proposal is a win-win-win for all three parties. Here's a more granular look at what is at stake and what can be achieved:

> Israel would meet its Paris Climate commitments to increase renewable energy capacity at cheapest cost and see regional cooperation strengthened. Jordan would achieve water security at cheapest cost through purchase of Israeli and Palestinian desalinated water and become a major exporter of green energy, to not only power

Mediterranean desalination plants, but also to sell enough solar energy to supply a substantial part of total regional energy consumption; and Palestine, in addition to becoming a water exporter to Jordan and perhaps the Negev in Israel, would become more independent from Israel to meet its water and energy needs.

What's holding all this up? Long standing rivalries, conflict, and mistrust between Israel, Palestine, and Jordan have continued to block a bioregional governance that would benefit all three parties and put the region on course to co-sharing their edge of the Mediterranean ecosystem. This first of a kind bioregional governance initiative in the Middle East came crashing down in October 2023 when Hamas terrorists invaded Israeli kibbutzes near the border of Gaza, killing 1,100 men, women, and children. The terrorist attack triggered a counter offensive with Israeli troops crossing into Gaza leading to thousands of civilian casualties – once again putting cooperation and bioregional governance on the backburner.

Whether all three parties – Israel, Palestine, and Jordan – and other fledgling joint efforts around collaborative governance can put the past to rest and make the transition from traditional geopolitics and "zero-sum games" to biosphere politics and "the network effect" that comes with sharing the responsibilities and opportunities that a bioregional governance offers is the big unknown in the Mediterranean region. Even though no country can any longer go it alone in the throes of a warming climate but will need to soften their sovereignty to include shared bioregional governance of their common ecosystems if our species is to survive and thrive, whether the turn will be taken and in time is very much an open question.

PART III

We Live on Planet Aqua and That Changes Everything

CHAPTER 7

Freeing the Waters

Freeing the waters is easier said than done. We need to remember that the defining signature of the past 6,000 years of human civilization is the domestication of the hydrosphere – capturing it, damming it, canalizing it, reorienting it, propertizing it, privatizing it, consuming it, profiting from it, depleting it, and poisoning it. It's been a fraught journey from the rise of the ancient hydraulic civilizations in the centuries before Christ to the hydro-powered superdams, artificial reservoirs, canals, and ports of the twenty-first century. The waters have been thoroughly repurposed to serve our species' conveniences. But, more often than not, these hydraulic infrastructures that manage water for our well-being are at the expense of the millions of other species who also depend on the availability of water.

The harnessing of the hydrosphere in its many modalities has largely determined the temporal and spatial orientation of whole societies and, in turn, the economic, social, political, and civic distinctiveness that has characterized each culture across recorded history. Equally so, it's fair to say, when looking back at the historical record, that the design and engineering of hydraulic infrastructure have at least partially fated the entropic bill that led to the demise and even collapse of many societies.

But, unlike in the past, this time around, the entropic bill brought on by the harnessing of the waters to a fossil fuel-based industrial revolution – the water–energy nexus – has eclipsed localities, regions, and even continents, taking the planet into the sixth extinction of life.

In a very real sense, the hydrosphere is now freeing itself – spasming in ways we could never have even imagined a half century ago. The waters are breaking loose with the steady rise of the Earth's temperature from the emission of global warming gases and altering the way water cycles the Earth, with seismic effects that our species can little fathom, nor respond to in kind.

Fortunately, there is the beginning of a conversation around the need to harness the human spirit and mobilize a sustained collective effort to assist in the freeing of the waters and allowing the hydrosphere to run its course as it self-evolves in an effort to accommodate a warming planet. Although this initial response is admirable and imperative and not without significance, it will be more in the way of our species finding ways of letting go of the many infrastructural restraints we have imposed on the hydrosphere over the decades and centuries – especially in the past two hundred years of a fossil fuel-based industrial era.

In the opening decade of the twentieth century, when both of my parents were born, much of the planet was still defined as wild and untouched by civilization. Today, less than 19% of the planet is still wilderness, as human development has diminished and in some cases eliminated ecosystems across the world.[1] But it's now the hydrological cycle that is taking the lead in deconstructing the hydraulic infrastructure and rewilding the Earth in wholly new ways, leaving our species to ponder the question of whether we will be resilient enough to go along for the ride and adapt to the new nature that is emerging.

Opening the Floodgates

In 2011–2012, the U.S. National Park Service detonated the Glines Canyon Dam and the Elwha Dam on the Elwha River in Washington State. The 210-foot Glines Canyon Dam, at the time, was the largest dam removal in the world. The Elwha Dam, built in 1914 to generate electricity, was heralded for electrifying the region but was not without critics.[2] The river was the richest salmon river on the Olympic Peninsula and the main source of livelihood for the Lower Elwha Klallam Tribe. The dam blocked salmon migration up the Elwha, drastically affecting the livelihood of local fishermen. A century later, with growing public interest in rewilding rivers and bringing back native wildlife, the political mood had shifted with the widespread support of freeing the waters and letting the salmon run again.

Nor is the Washington State decision to demolish the dams and assist in the freeing of the waters a one-off event. Increasing public concern over the loss of natural habitats and the diminishment of wildlife and the wilderness itself, especially coming on the heels of a catastrophic change of the climate from global warming emissions, has energized a younger generation to champion the hydrosphere. In the boroughs of New York City, the heart of urban America, citizen scientists have uncovered old maps that show brooks, streams, and rivers flowing all across the island. Using the maps as a guide, enthusiasts have discovered that under the pavement there's an intricate system of native streams that still exist. The goal is to free the waters under the surface and let them run their natural course, while readapting the urban landscape to adjust to their flows. For example, when the citizens plant trees, they are looking to place them where there is a natural flow of water underneath the pavement, enhancing their ability to grow. Likewise, homeowners and apartment dwellers who experience constantly flooded basements can search the ancient maps to find out if their buildings lie above a still-functioning stream and construct an appropriate drainage system to keep their buildings dry.[3]

These efforts to free the waters of New York City and rebuild the urban infrastructure to adapt to the currents and flows of its long-hidden rivers and streams took on a more urgent note after Hurricane Sandy swept through New York on October 29, 2012, flooding subway systems, tunnels, roads, and buildings, and closing the New York Stock Exchange for two days. Thousands of New Yorkers had to be evacuated for nearly a week in the storm's aftermath. Forty-three people died in the storm, which caused an estimated damage of $19 billion dollars.[4] Other cities across the U.S. are following New York City's lead, including Dubuque, Kalamazoo, and San Francisco, uncovering ancient streams underneath the pavement and freeing the waters with the objective of reintegrating the urban environment with its ancient rivers.

Seventy percent of the Earth is water, but only three percent is fresh water, and 0.1% of that fresh water is available, making it an extremely rare resource.[5] Unfortunately, in the course of development during the industrial era, urban and later suburban communities were built over once vibrant floodplains, since drained, dammed, and diverted. The loss of valuable water came with a concomitant loss of wildlife. In Great Britain alone, to use one example, it is estimated that 90% of its once lush wetlands have disappeared, along with much of its wildlife, with the rise of an urban industrial landscape.[6]

Across the U.K., efforts are underway to free the waters and reconnect rivers with their natural floodplains, restore wildlife habitats, and decommission and demolish dams with the goal of learning to adapt to the waters rather than adapt the waters to human development. Nor are these efforts trivial. It is estimated that one in six properties in the U.K. alone are at risk of flooding, and with the hydrological cycle ramping up record floods, climate threats to the building stock will worsen as the Earth's temperature continues to rise.[7] Freeing the waters and learning to adapt to a rapidly changing hydrosphere is no longer a whim but a prerequisite for survival.

In Britain, as elsewhere, citizen scientists and volunteers working with marine biologists and local governments are rewilding fish nurseries and salt marshes, regenerating kelp forests, restoring native oyster stocks, saving seagrass habitats and engaging in other projects that can soak up carbon emissions, reduce flooding, cleanse marshes and marine areas, and bring back native species.[8] Initiatives are even underway in releasing once omnipresent beavers back into the wild to help recover ecosystems, provide natural flood protection, and slow the flow of pollutants downstream.[9]

While the sequestration of the waters on land masses has seriously damaged the lithosphere, endangering the future of all the land-based species, the great oceans were not to be spared. Over eons of history, the idea of capturing and domesticating the oceans would have seemed beyond human imagination. There was little thought of seizing the oceanic expanse and enclosing it for commerce and trade. That changed in the late fifteenth century with the great seafaring voyages to find and colonize other continents and establish new potential markets. At the time, Spain and Portugal were the undisputed sea powers and were constantly engaged in open combat on the oceans in their rush to claim new lands for their respective kingdoms. They decided that it would be more practical to put an end to open warfare and instead divide up the world's oceans between them.

In 1494, the two countries signed the Treaty of Tordesillas, dividing the world's oceans into sovereign spaces "between the north and south poles that ran 370 leagues west of the Cape Verde Islands."[10] Spain gained exclusive sovereignty of the ocean west of the demarcation line, including the Gulf of Mexico and the Pacific Ocean, while Portugal was granted everything east of the demarcation line, including the Atlantic and Indian Oceans. Sir Walter Raleigh, the English explorer, summed up the significance of the arrangement, pointing out that "whosoever commands the sea, commands the trade: whosoever commands the trade of the world, commands the riches of the world, and consequently, the world itself."[11] Raleigh's off-handed comment would become an intricate part of geopolitics in the nineteenth and twentieth centuries, an era in which imperial powers fought one another to command the open seas and commerce and trade on fledgling global markets.

The rivalry notwithstanding, it soon became obvious that no single nation could hope to exercise sovereignty over the expanse of the oceans. So countries began to nibble away at regions of the oceans extending out from their coastal waters. The Italians made claim to a hundred-mile zone from their coast out to sea, with the justification that the distance represents how far a ship could sail in two days. Other countries claimed sovereignty on open waters as far as the eye can see. Even more ambitious, a few nations suggested extending sovereignty over the oceans as far as the telescope could see. The Dutch upped the ante, claiming that sovereignty should extend as far out from the coast as a cannonball could be fired. That new demarcation became the standard until the eve of World War II. The American rear-admiral and naval strategist, Alfred Thayer Mahan, summed up the prevailing zeitgeist among the world's premier governing powers, suggesting that "control of the seas by maritime commerce and naval supremacy means predominant influence in the world . . . [and] is the chief among the merely material elements in the power and prosperity of nations."[12]

All of these initiatives to establish suzerainty over parts of the global oceanic commons created more confusion and conflict, tiring every wannabe sea-power nation. By the late twentieth century, widespread agreement began to gel over developing a more coherent plan to divvy up the oceans. In 1982, the United Nations established the Law of the Sea Convention, granting each nation sovereignty twelve miles out to sea. Equally important, the treaty also included the granting of exclusive economic zones (EEZs) up to two hundred nautical miles offshore, giving each country "sovereign rights for the purpose of exploring and exploiting, conserving and managing" the living and nonliving resources of the oceans, seabeds, and subsoil. This unprecedented give-away allowed coastal nations to enclose much of the planet's oceans, containing 90% of the marine fisheries and 87% of the offshore oil and gas reserves along EEZs.[13]

The most sought-after prize in recent decades has been the enormous wealth in oil and gas at the bottom of the oceans and seas. The oceanic seabed is also a treasure trove of valuable minerals and metals, including manganese, copper, aluminum, cobalt, tin, lithium, uranium, and boron. Vast stretches of the ocean floor have been doled out to nations, enclosing the last great commons and much of the hydrosphere that governs Planet Aqua.

Open-sea fishing is still another egregious expropriation of the ocean's bounty. It's no secret that overfishing has depleted fish stocks virtually everywhere. The fishing industry uses high technology monitoring, including satellite surveillance, radar, sonar and seabed mapping to locate deep-sea fishing fields, transforming the fishing industry into the "strip miners" of the ocean depths. The key players use gigantic trawler ships weighing as much as fourteen thousand tons, and stretching the length of football fields. The trawlers are literally floating factories that catch, kill, process, and package their haul on board. These trawlers lay out upwards of "eighty miles of submerged longlines or forty-mile drift nets."[14] The hyper-efficiencies of high-tech ocean fishing have so devastated global fishing stocks that about one-third of fishing revenues now come from government subsidies just to keep the industry viable.[15]

The commercial exploitation of the great oceans has put them on life support. Unless the ocean waters are freed of geopolitical exploitation they will die, and so too life on Earth. That prospect, once just a possibility, is now within range. The big picture answer is to turn large swaths of the oceans into Marine Protected Reserves to bolster the resilience of ocean ecosystems. In 2016, the International Union for the Conservation of Nature adopted a resolution calling for the placing of 30% of the oceans under protection by 2030 to avert large-scale extinction of life. An increasing number of countries have set aside areas of the oceans within their governing jurisdictions as protected and unavailable for commercial exploitation, including the U.S., U.K., Mexico, and Chile.[16] Still, less than five percent of the oceans have formally been placed under protection, and only two percent are highly protected.[17]

That is now likely to change. In 2023, the United Nations adopted an expansive treaty to protect the high seas and conserve biodiversity, along with a comprehensive plan to establish large-scale marine protected areas.[18] The goal is to conserve and manage 30% of the sea and coastal areas and includes guidelines for conducting environmental impact assessments to assure the safeguarding of marine ecosystems. The treaty will come into effect after ratification by sixty nations. Already, the European Union and its 27 member states have signed the treaty.

While the great oceans were reduced to property and are now failing, the scarce fresh water on the planet has also become propertized and privatized in a global marketplace managed by a handful of corporations. Until recently, the fresh water of the planet had always been publicly administered on an open commons. Suddenly, in just the last half century, the fresh water of the planet was seized by global companies and made over into a commodity in the marketplace.

It all began in the early 1980s when the United Kingdom, the United States, and other nations, in tandem with global governing institutions, including the World Bank, the International Monetary Fund (IMF), the Organization for Economic Co-operation and Development (OECD), and the United Nations, etc. began promoting the transfer of public water utilities to the private sector. The World Trade Organization (WTO) has classified water as a "tradable commodity" and a "commercial good," "service," and "investment," and locked in regulations that constrain governments that attempt to prevent the private sector from controlling the global water business in the marketplace.[19] The rationale was that letting the private sector take over water utilities would ensure the most efficient operating practices to deliver the best market prices, all to the advantage of consumers.

The World Bank and other lending institutions championed so-called public–private partnerships (PPPs) encouraging governments to lease their water infrastructure to private sector companies to manage. What the World Bank, the OECD, the WTO, and other global institutions failed to see is that there is little incentive on the part of private companies to either upgrade the public infrastructures and services they are managing or reduce prices. At least with markets where consumers have a choice of providers and can switch to competitors who can offer a better price and improved service, public infrastructure – roads, airports, et cetera – are natural monopolies and freed up from competition. The sad reality is that public–private partnerships with long-term leases often encourage companies to engage in "asset stripping" – that is, not adding improvements to the infrastructure and services, knowing that their users have little or no alternative to the service rendered.

Unlike publicly administered infrastructure services, market-based companies are constrained from the get-go in continually maintaining less improving services for the simple reason that they have to show a steady gain in their revenue streams and profit, although their consumer base remains relatively flat. In other words, the potential market is often tapped from the start. The result is continuous asset stripping in order to save on costs and ensure that profits flow. This is particularly the case with water and sanitation systems, where the most impoverished communities have no choice but to accept whatever conditions private companies impose.

In the early years of the privatization of water, the World Bank encouraged public–private partnerships through generous loans to governments, and through its private sector arm, the International Finance Corporation, whose mission, in part, is to invest in privatization projects. Even as evidence mounted on the shortcomings of the privatization of water, the World Bank continued to finance its privatization. The privatization process has yet to be quelled. Ten global companies now dominate the worldwide water utility services market.[20] These giant global corporations push a privatization agenda, benefiting from generous government incentives and subsidies, while raking in massive profits by charging high prices for water while risking reducing the quality of water services.

A study conducted on water services in the United States found that industry-owned utilities typically "charge 59% more for water service than local government utilities" and "charge 63% more for sewer service than local government utilities." Moreover, privatization "can increase the cost of financing a water project by 50 to 150%." The study reviewed eighteen municipalities that ended contracts with private companies and found that "public operations averaged 21% cheaper than private operations for water and sewer services."[21]

Harvesting the Hydrosphere

There is a lingering public misconception that a warming climate means we are running out of water. While the increasingly heavy rainstorms and floods are acknowledged, they are often treated as separate phenomena from the droughts. What's missing is that the planet is not running out of fresh water but that a rewilding hydrosphere brought on by a warming climate is changing both the seasonal timing of the rains and the intensity and duration of the waters falling from the sky. The problem is that the global hydraulic civilization is locked into a water cycle cued to a temperate climate that no longer exists and, as a result, is not able to deliver the waters for human consumption, industrial use, and irrigation of agricultural lands when and where it's needed.

Researchers at the Massachusetts Institute of Technology (MIT) back in 2014 addressed the issue in a report entitled *The Energy and Climate Outlook*. The authors introduced a controversial scenario in the public discussion by suggesting that climate change might actually increase the availability of global fresh water by about 15%. John Reilly, the co-director of the Joint Program on the Science and Policy of Global Change at MIT, noted that "all climate models predict a speed-up of the hydrological cycle with warmer temperatures." The warmer temperatures mean "faster evaporation, more moisture in the atmosphere, and more rainfall."[22] The report projects an increase flow of fresh water of around 15% globally by the latter decades of the twenty-first century, although the researchers cautioned that human use worldwide is likely to increase by 19%, in agricultural, industrial, and domestic uses.[23]

The problem again is when and where and also how much water will be distributed across the planet's hydrosphere. The current climate models generally agree that we are looking at more precipitation poleward and drier conditions in the mid latitudes and subtropics. The bottom line, says Reilly, is that "water stress, or not, is very much a function of precipitation in the right place at the right time, and in the right form."[24] For example, heavy rain in the fall after a harvest is of little assistance to farmers. The melting of snowpacks too early in the winter months means sufficient water won't be flowing later on in the spring and summer months for the irrigation of the crops. The scientists remind us that snowpacks, particularly from the mountains, are nature's water storage system, and their proper seasonal melting over the past 10,000 years has conditioned agriculture everywhere. Reilly concludes with the observation that "more rain is likely to come in heavier downpours, with longer periods in between," which means more flooding, runoffs, and droughts.[25] But again, it's not only a question of how much water falls during the passing of the seasons but when and where.

The long form answer here is to reset our relationship to the hydrosphere from adapting the waters to our species to adapting our species to the waters. A slew of initiatives along these lines is emerging and beginning to scale with catchy titles like "Slow Waters," "Sponge Cities," "Nature Based Systems," and "Green Infrastructure." What they all have in common is a paradigm shift from an overly centralized and hyperefficient hydraulic approach to sequestering the more predictable flow of the waters during the temperate climate of the Holocene era to a far more adaptive approach to adjusting to a rewilding hydrosphere in the Anthropocene. Or, to put it in more of a philosophical frame, these new approaches to adapting to the hydrosphere shift the balance from management of the waters to stewardship of the waters . . . going with the flow instead of directing the flow of the Earth's hydrosphere.

This distributed approach to adapting to the waters is less bound by centralized command and control exercised from the top down by nation states and administered and managed by national governments and corporate bureaucracies, all circumscribed by state political borders and more bound to hands-on engagement by local populations enmeshed in their own ecosystems in bioregional forms of governance. This intimate resetting of our species' sense of time, space, and attachment bound to one's immediate biosphere reintroduces the ancient form of "commons governance" but in a more politically and technologically advanced fashion befitting a twenty-first-century understanding of the world we inhabit.

This new extension of governance is trending. That doesn't mean that nation states are going to suddenly disappear, but that much of the heavy lifting in learning how to live with a warming climate and a rewilding hydrosphere is best served by the active participation of our species in the ecosystems where we live out our lives along with our fellow species. Consider the idea of "slow water," which has evolved from two earlier movements that spread across the world over the past two generations – slow foods and slow cities. Each of these resets are about letting go of the hyperefficient extraction, commodification, propertization, and consumption of the ecosystems that sustain life, and relearning how to adapt at the local level to the processes, patterns, rhythms, and timetables set by nature.

Erica Gies, a journalist and National Geographic Explorer who writes on water issues and climate change and coined the term "slow water," makes the point that water in its natural state is not always pouring down and rushing across the land surface but also has its "slow stages," for example, when seeping down into the soil or settling on wetlands or nestling into groundwater caverns, and she says that this is "where the magic happens" by "providing habitat and food for many forms of life above and below." Gies suggests that "the key to greater resilience . . . is to find ways to let water be water, to reclaim space for it to interact with the land."[26] The issue is that our global hydraulic civilization is designed and engineered to sequester water quickly, store it in artificial reservoirs, and pump it across pipes to its destination, to irrigate the crops or over dams to generate electricity or into homes and businesses, to draw water and then quickly recycle the wastewater back again to be repurified and sent back once more across the entire matrix.

To complicate matters, the buildout of complex hydraulic systems led to growing populations living together in ever-more dense urban and later suburban communities, all built over former wetlands, rivers, and streams – so the rains have nowhere to go, causing massive flash floods to flow down onto impermeable cement and asphalt surfaces, which prevents the waters from permeating into the soil and from providing the nourishment the soil requires to sustain the ecosystems of the Earth. To take one example that can be multiplied over and over again in urban and suburban communities, Gies reports that in many of China's dense urban sprawls, less than 20% of precipitation running off buildings and pavement soaks into the soil but, rather, finds its way into drains and pipes that take it away. In Beijing alone, where pumped groundwater long supplied its growing number of residents, the water table has been dropping by a meter every year, with consequences that until recently were ignored. Letting go of the waters to find their natural flow is the way forward for our species' realignment to the planet.[27]

Experiments are going on everywhere introducing new ecosystem-friendly ways to free the waters, but most are pilots and few have begun to scale. Engineers, urban planners, and landscape architects are introducing bioswales and rain gardens to free the waters. A bioswale is a long hollow channel or trench that is filled with native grasses, shrubs and flowers grown in soil and mulch, accompanied by layers of stones to "slow down" rainwater and simultaneously filter out pollutants, including petrochemical fertilizers, motor oil, and litter. Rain gardens perform a similar function but in a different way. While bioswales slow down rainwater through a curving or linear path, rain gardens are designed in such a way as to "capture, store, and infiltrate rainwater in a bowl shape."[28] Permeable pavement is also beginning to replace conventional pavement in cities and suburbs. These new pavements can be asphalt, porous concrete, interlocking pavers, and plastic grid pavers, and allow the rain and melted snow to flow into the underlying soil.

Green rooftop gardens can also indirectly slow down water. These elevated gardens, replete with vegetation, "provide shade, remove heat from the air, and reduce temperatures of the roof surface." Green roofs are becoming increasingly popular and can dramatically reduce the temperature on rooftops and, if scaled across urban areas, "can reduce city-wide ambient temperatures by up to five degrees Fahrenheit," slowing down evapotranspiration, allowing water to sink into the soil, while reducing electricity demand.[29]

A more integrated and scaled approach to slowing down the waters and letting them permeate into the soil or be stored in underground cisterns and artificial aquifers beneath urban sprawls has captured the public imagination. It's called "sponge cities." The concept is the brainchild of Kongjian Yu, a Chinese architect and urbanist. Yu is the founder of the Peking University College of Architecture and Landscape Architecture. The "sponge city approach" is based on bringing back natural waters to urban areas. Landscape architects remind us that many of the great cities of the world emerged on and around lakes, rivers, and wetlands over centuries and millennia. These increasingly dense urban milieus paved over the waters and relied exclusively on pumped water brought in via centralized hydraulic infrastructures, often from outlying regions. Now, even these ancient waters – especially the great rivers and lakes of the world – are diminishing from overextraction and a warming climate, throwing urban communities into a crisis of survival. The sponge city is an attempt to prevent flooding by slowing down falling rain via the introduction of natural landscapes throughout the urban sprawl, letting it seep down into the ground to replenish local groundwater or be piped into underground water tanks for storage, and later tapped when needed.

The Turkish city of Izmir is a growing urban sprawl with a population of over three million people, and climbing. The city is particularly vulnerable to severe "cloudbursts." The city is also one of the fast-growing metropolitan areas in the Mediterranean and has doubled in population over the past thirty years. Much of the urbanization process spread beyond the traditional city contours and into agricultural lands, resulting in extensive "soil sealing," leaving the expansive urban area prone to massive flooding across its paved roads and byways. New studies show that entire parts of the city rest on ground surfaces that are 75% to 100% sealed, meaning that increasingly virulent downpours are unable to sink into the ground and down into the groundwater underneath but race through city streets, leaving homes and businesses flooded and large parts of the city underwater for brief periods of time.[30]

The Izmir city planners teamed up with the Izmir Institute of Technology and together created a digital model, mapping out in granular detail the nature and type of sponge intervention required in all the city's quadrants. The resulting guidance document established a roadmap of site-specific natural landscapes located in the city to slow the waters, along with catchments to save the waters for use in non-raining periods, giving the city a leg-up on building resilience throughout its urban infrastructure.

Some critics agree that there is value in providing citywide wetlands and parks to retain storm water but argue that they may not be able to handle the kind of superstorms coming our way in a warming world. Yu, however, points out that Chinese cities are required to retain 30% of the city as greenspace and an additional 30% dedicated to community space, which ought to be more than sufficient to create more ponds and water-absorbing parks that can capture large amounts of falling water.[31]

The nub of climate change is that a warming air in the atmosphere sucks up more moisture from the ground, all of which translates into heavier snowfalls in the winter, torrential floods in the spring, prolonged droughts, heatwaves, and wildfires in the summer, and devastating hurricanes in the autumn months. The changes in where, when, and how much fresh water falls to Earth mean that rainwater needs to be "stored" to be available in the months of severe drought, extreme heatwaves, and cataclysmic wildfires.

Harvesting rainwater, once considered an outdated ancient practice, has suddenly become a priority, particularly in arid and semi-arid regions of the world. And, if we think it's only the poorest village communities in developing countries where the ancient technology is making a comeback, not so. From Jordanian villages and towns to the world's high-tech cities like Las Vegas, harvesting rainwater by combining traditional methods, IoT sensing, and algorithm technologies is becoming a player. Within the next century, hundreds of millions of cisterns will be operational, some stand-alone and others distributively connected, allowing harvested water to be shared in local neighborhoods in any given community.

In Jordan, dating back to Roman times and the Byzantine era, cisterns have been used to harvest water. Many have remained in use to the present day, while other long dormant cisterns are being given a second life, refitted for use in a climate-changing environment where precious water has to be captured when it rains during the winter and early spring months, to be used when the land is parched in the summer months and the early fall season. Many of the tanks are olla-shaped and built below a limestone catchment. Villagers use the harvested water for drinking and cooking, often supplementing it with trucked-in water used for bathing and irrigation.

Using jackhammers, workers excavate the underground tank area and plaster it using either cement or clay – the latter being preferable, since cement comes with a high carbon footprint. With new buildings, downspouts on the roof direct the water down into the underground tanks, the latter equipped with a pump that can direct the harvested water back up into the home's plumbing infrastructure. These stand-alone water harvesting systems are complemented by more extensive scaled water harvesting structures. For example, in Jordan, Saint George's, a nineteenth-century Greek Orthodox church in Madaba, constructed over a fourth-century cistern, includes a latticework of underground channels carved out over time, which are collecting runoff from water seeping down under the worn cobblestone streets in the urban commune and elsewhere.[32]

Between 2018 and 2020, the United States Agency for International Development Small Projects Assistance program (USAID SPA) funded Peace Corps volunteers and local communities across nine municipalities in four states of Mexico to install water harvesting systems in 68 homes and 23 schools and community centers, with a total capacity of 1,633,330 liters of rainwater.[33] Local engineers working alongside Peace Corps volunteers and neighborhood work crews installed 12,000-liter cisterns in homes and large-scale 50,000-liter cisterns in schools. This type of effort is being duplicated around the world, where entire communities are beginning to collect water during the rainy season in the spring to be used during the drought-prone seasons later in the year.

Among the more ambitious efforts is the One Million Cisterns for the Sahel, a plan sponsored by the UN Food and Agricultural Organization (FAO). Water harvesting and storage systems, says the FAO, are being installed across seven countries in the Sahel – Senegal, Gambia, Cabo Verde, the Niger, Burkina Faso, Chad, and Mali.[34] The effort is directed to the most vulnerable rural communities in these arid and semi-arid regions. All of the countries are plagued by massive floods, followed in close time by repeated droughts – and the impacts of these increasingly wild gyrations in the water cycle brought on by a warming climate, notes the FAO, "are devastating for the poorest rural households, who struggle with these shocks and see their vulnerability worsen."[35]

The objective of this ambitious scaled effort to introduce rainwater harvesting across seven countries, says the FAO, "is to enable millions of people in the Sahel to have access to safe drinking water, to enhance their family agriculture production to create a surplus, improve their food and nutrition security and strengthen their resilience," with the focus on women.[36] The program is designed to train local communities in the construction and management of the cisterns by way of "cash-for-work" activities, with particular attention to engaging women in the execution of these scaled water harvesting cisterns. This capacity building is an intimate part of the program to ensure that the task of deploying and managing this distributed approach to adapting to the freeing of the waters is a gender-neutral exercise – enabling local populations to oversee a hydrological commons across seven nations who share a common ecosystem in the Sahel. According to the FAO:

> Local communities are trained in the construction, use, and maintenance of cisterns, thus becoming qualified for civil construction works and infrastructure maintenance to enable income diversification and

> improved housing conditions. They are also trained on good water management. Training courses on adaptation to climate change in agriculture and agroecology are also organized in synergy with farmers' field school programs.[37]

Even the United States, the most highly industrialized and arguably technologically advanced nation in the world, is climbing onboard water harvesting. Rhode Island, Texas, and Virginia, for example, offer tax credits for equipment purchased for rainwater harvesting. Some states, however, have restrictions on the amount of rainwater that can be collected, and how it's collected.[38] Water harvesting, until recently, has been mostly restricted to non-potable uses – crop irrigation, gardens, and yards.

Another compelling reason why water will have to be stored when it rains, aside from climate-related seasonal changes in its availability, is the increasing risk of cybercriminal incursions and terrorist attacks aimed at crippling pumping stations and pipelines, closing off water in large urban areas, leaving millions of people stranded and at risk of severe dehydration and even death. Were the central water grid to be compromised, thus shutting down the main water system across entire regions, neighborhood water microgrids can be turned on locally to ensure that the flow of water is maintained until the main water line is restored.

A water internet replete with IoT sensors is just now beginning to be embedded in water reservoirs and in the pipelines that bring fresh water to consumers and remove wastewater, sending it back to treatment plants for repurification. The IoT sensors monitor pressure on the pipes, the wear and tear of the equipment, leakages across the system, and the change in water clarity and chemistry, using data and analytics to anticipate, intervene, and even remotely fix trouble spots along the line. Smart meters and sensor monitoring, in turn, provide "just in time" data on water flow, including the volume and time of usage, to manage water resources more effectively, from ensuring clean water distribution to recycling and purifying wastewater that can then be reused by consumers, thus saving water in a virtuous circular system. The embedding of a water internet throughout our water systems becomes ever more pressing when we consider the fact that, for example, in the United States, nearly six billion gallons of treated water are wasted every day because of leaking pipes, metering inaccuracies, and other errors, according to the American Society of Civil Engineers.[39] Other countries face similar water losses.

The Pacific Northwest National Laboratory operated by Battelle for the U.S. Department of Energy prepared a detailed study in 2021 on the need to introduce a secondary distributed backup water microgrid infrastructure across the United States that allows localities and even neighborhoods to meet "pressing water demands," emphasizing the need to provide water storage in a similar way to that in which batteries and fuel cells store electricity in microgrids.

Water microgrids can even treat their own locally sourced water onsite for both potable and non-potable use. The Omega Center for Sustainable Living – an institute that teaches courses and hosts conferences on sustainability and resilience – pioneered one of the early onsite water purification systems that mimic the way nature purifies water. At Omega, water is drawn from an underground aquifer and pumped uphill to a cistern to create water pressure. The pure water then flows down in pipes and into the campus's many classrooms, dormitories, and dining hall, where it is used for drinking, bathing, cleaning, cooking, and toilets. The used water then flows back into an eco-machine to be purified and from there back into the aquifer where it originated. The ecomachine can process upwards of 52,000 gallons of water per day. This particular system using net zero energy is a closed loop hydrological cycle as close to nature's recycling system as possible. The purification system, which was developed by John Todd, a pioneer in developing circular systems patterned after those in nature, utilizes the same purification process found in estuaries – nature's "own water filtration system."[40]

In this system, wastewater is pumped back into "highly oxygenated aerated lagoons . . . [where] plants, fungi, algae, snails, and other micro-organisms . . . are busy converting ammonia into nitrates and toxins into base elements." From the aerated lagoons, the water is sent on its way to a recirculating sand filter where both the "sand and micro-organisms absorb and digest any remaining particulates and small amounts of nitrates" after which it's pumped into two fields under the Omega Institute parking lot, where it disperses into the groundwater below the surface and eventually trickles down into the aquifer 250 to 300 feet below the Omega campus. The purified water is then drawn, pumped up to the buildings, and used again for drinking water, cooking, cleaning, showers, and toilets in a closed loop.[41]

The Omega Center's onsite water repurification system was among the first of its kind to demonstrate the viability of recycling water over and over again onsite in a self-contained circular loop. In recent years, the onsite recycling of water has begun to scale with the introduction of new compact stand-alone water repurification systems that can be purchased and installed in basements of homes, hotels, commercial buildings, factories, and tech parks. The size of commercial refrigerators, these sleek appliances are attached to pipes throughout the building, which collect "gray water" from showers, sinks, and washing machines. The gray water is then sent back to the basement, where it is cleaned by using membrane filtration, ultraviolet light, and chlorine and sent back upstairs for non-potable uses with virtually no losses. Peter Fiske, the Executive Director of the National Alliance for Water Innovation at the Lawrence Berkeley National Laboratory says, "we now have technologies to enable us to process and reuse water over and over at the scale of a city, a campus, and even individual homes."[42]

Distributed and decentralized onsite water systems like solar and wind are already beginning to scale and experience their own plummeting exponential cost curve. The market is vast and potentially covers upwards of hundreds of millions of buildings. In a half century from now, distributed and decentralized water systems are likely to be ever-present in buildings, alongside the sinks, showers, and washing machines that they service. Local governments are quickly stepping forward to mandate the installation of decentralized water systems. In 2015, San Francisco began requiring that "all new buildings of more than 100,000 square feet have onsite [water] recycling systems."[43]

The Salesforce Tower, a new 61-story hotel, office, and residential tower that opened in 2018 and is the tallest building in San Francisco, includes an onsite water recycling system that takes in shower, sink, and sewage water, cleaning 30,000 gallons of wastewater each day to be reused for toilet flushing and irrigation, with an estimated saving of 7.8 million gallons of water per year – that's equivalent to the water used in 16,000 homes in San Francisco.[44] A similar project currently under construction in Brooklyn, New York, by Domino Refining will recycle 400,000 gallons of water each day.[45]

At present, the purification system only pertains to black water from toilets, dishwashers, and kitchen sinks, and gray water from washing machines, showers, and bathtubs. The recycling of non-potable water for flushed toilets and washed clothes reduces water demand by 40%, and the recycling of water for showers reduces demand by another 20%.[46]

The San Francisco Public Utilities Commission estimates that by 2040, onsite water reuse projects will "save 1.3 million gallons of potable water each day." Hydrologists estimate that the repurification of potable water at-scale in decentralized systems is only five or ten years away.[47] Installing water recycling systems adds approximately six percent to the cost of a home and twelve percent for multi-dwelling buildings. The return on investment is around seven years, after which residents enjoy an appreciable savings on water and sewage costs.[48] And these upfront costs are expected to drop appreciably with economies of scale.

Within the next several decades, decentralized water systems will be ubiquitous. Looking down the road, Fiske says that "new buildings and neighborhoods . . . may someday no longer need to hook up to sewer lines and water supplies. People will be able to build without regard to connections to water infrastructure, simply by using the same water again and again in a virtually closed loop," adding that "the water that falls on the roof in most places in the world will be enough to sustain a home." The decentralization of water, according to one recent study, could save as much as 75% of water demand.[49] Like the local harvesting of the sun and wind, both of which are currently scaling all across the world, the harvesting of water on hundreds of millions of rooftops in neighborhoods decentralizes and democratizes this most basic life source on which our species relies to sustain life.

Although all of these highly distributed efforts together will give cities some breathing room and a bit of an edge in adapting to a rewilding hydrosphere, they beg the question of whether our species' mega-cities, which overlay a massive hydraulic infrastructure, are the appropriate habitats to weather the atmospheric rivers, floods, droughts, heatwaves, wildfires, and hurricanes coming on. Dense urban hydraulic civilization may have been an appropriate fit for a temperate climate but not for a planet that is heating up.

Or, to put it more bluntly, if the dense urban hydraulic civilizations were the dominant way our species came to organize life over the six thousand years of a temperate climate, and whose core mission was to adapt nature and the Earth's hydrosphere to our species, how can we expect that the same infrastructure and accompanying practices and worldview will be a good fit in extricating our species from the planetary crisis this ancient worldview spawned? Perhaps these new environmentally friendly water initiatives will be looked back on as waystations on the way to rethinking how to live out our lives by adapting our species to nature rather than the other way around.

At a moment when the human family is despairing of the future, the Age of Resilience brings with it a new and powerful narrative, which, if widely embraced, might lay the basis for a radically different future that can bring us back into nature's fold, giving life a second chance to flourish on Earth.

CHAPTER 8

The Great Migration and the Rise of Ephemeral Society

If readapting to the waters rather than continuing to force the waters to adapt to our species seems a bit much to imagine, we need to recall that our species has been there before. Our human family is nomadic by nature. Now, another great climate migration has begun, marking the beginning of a new nomadism as entire segments of the population around the world search for more temperate weather regimes in which to live. A stream of new demographic studies and reports suggest that over the next 45 years one in twelve Americans will likely flee the southern half of the United States and head toward the mountain west and northwest. Millions of others will pick up their belongings from their drought ridden environments, heatwaves, and fire-sensitive southwestern states and migrate to the Great Lakes region.[1]

These migrations, now a trickle, will likely become a wave over the next several decades. The Midwest farm belt, the breadbasket of the world, is already facing severe climate-related spring floods and summer droughts year-after-year, turning America's once lush crop lands into giant lakes at the beginning of the planting season and parched soil toward the end of the season. Farmers, hard hit, are filing for bankruptcy. Meanwhile, consumers are facing spiraling inflation, in large part caused by higher prices for food at supermarkets, and are having to scale back on groceries – all because of a fast-changing hydrological cycle in a warming climate.

After 10,000 years of relatively temperate climate in which our species forced nature's primary agency – the hydrosphere – to adapt to human society, we will be forced to make a 180-degree turn around and once again find new ways to adapt to the earth's hydrological cycle along with our fellow creatures. But this time around and equipped with a more sophisticated scientific and phenomenological understanding of the workings of the hydrosphere and how it affects the Earth's other three agencies – the lithosphere, the atmosphere, and the biosphere – we can begin to reorder our technological prowess from one of exploitation to stewardship, finding ways to engage the blue-marble as kindred spirits.

Much is in doubt as we head into the greatest change in the earth's climate in more than 56 million years.[2] Yet there is reason to believe that our species, and hopefully many of our evolutionary relatives, will survive and even flourish in totally new ways of which we can only imagine at this crossroad in the human journey. Blessed with extraordinary adaptive traits, we traversed the Earth, crossed the oceans and great land masses, surviving ice ages and warming thaws. Thinking of the future ahead as a threshold instead of a death knell will likely be the powerful tonic that gets us through to the other side.

Our human family is already experiencing the first wave of what will assuredly be the greatest mass migration of our species in history. The new nomadism will bring with it a change in our spatial and temporal orientation as our species, long used to a sedentary life across the generations, will be forced to migrate to more resilient regions to escape the severe climate-related disruptions triggered by the earth's supercharged hydrological cycle.

The upcoming human migration will test our ingenuity. If the Holocene was characterized by long periods of sedentary life and short periods of migratory life, the Anthropocene is likely to reverse the dynamic, with shorter periods of sedentary life and longer periods of migration, in line with the pace set by the hydrosphere. That so, we are beginning to see the early shoots of what might well be the rise of an ephemeral society. Pop-up ephemeral cities of tens and even hundreds of thousands of migrants are already here, but only roughly defined, and they operate more like refugee camps. But, as human migration becomes a way of life and not a temporary interruption, pop-up cities along migratory corridors will become more defined and refined and more suitable to a spartan, but good quality of life and likely be administered collectively under the aegis of "commons governance." Within the next five decades we can expect to see the erection of ephemeral cities that will also be mobile and that can maintain stride with changing migratory patterns and shifting travel corridors, equipped with state-of-the-art structures that can be assembled, disassembled, moved, and reassembled along with the flow of human traffic, as our species acclimates to climate disruptions.

The new reset in how our species might flourish in an ephemeral society punctuated by a global warming planet and a rewilding hydrosphere is manifesting itself in a fundamental rethinking in the way our human family lives in time and navigates space. The first inkling of this temporal and spatial reorientation comes in the way we are beginning to reconceive habitats and attachment to place. The conventional notion of what constitutes a safe haven has been thrown to the winds or more accurately thrown to the waters in a highly unpredictable climate-changing world. The ancient ontological and philosophical struggle between "being" and "becoming" that viewed the former as real and the latter as mercurial or even illusionary is reversing, as we venture further into a volatile twenty-first century.

The idea of reality as made up of objects, structures, and forms that are predictable, atemporal, and passive is quickly slipping by the wayside in philosophical circles and even within the scientific community, where the talk is increasingly drifting away from capturing "things" and sequestering passive matter and treating all phenomena as objects to a new understanding of reality as self-evolving processes, patterns, and flows.

The Renaissance of the Ephemeral Arts and the Reset of Time and Space

The arts are generally a forerunner in reorienting our changing relationship to time and space over history. The last great reset in the arts dates to the Medieval era and the Italian Renaissance between 1340 and 1550 – a period that was marked by the reimagining of time and space. The temporal shift began with the Benedictines, a monastic order founded in 529 by Benedict of Monte Cassino. The Benedictines were fiercely committed to strenuous manual labor and strict religious observance. Their guiding principle was "idleness is the enemy of the soul." Manual labor for the Benedictines was a form of penitence in order to secure their eternal salvation. St. Benedict cautioned the brothers that "if we would escape the pains of hell and reach eternal life, then must we – while there is still time, while we are in this body and can fulfil all these things by the light of this life – hasten to do now what may profit us for eternity."[3]

Scholars point out that the Benedictines were likely the first to think of time as "a scarce resource" and, because time belonged to God, each of the brothers had to use every moment of every waking hour paying homage to the Lord. Each day was dedicated to relentless organized activity. Specific times were allocated for prayer, labor, study of scripture, eating, bathing, sleeping, etc.[4] Specific times were even established for the most mundane chores, including head shaving, bloodletting, and mattress refilling, etc. No time was left unfulfilled.

Anthropologists suggest that the Benedictines were likely the first cohort in history to adhere to schedules – a practice that has become a way of life and a signature of the natural order of things. For that reason, they are often categorized as "the first 'professionals' of Western civilization."[5] Despite the fanatical zeal to be on time, all the time, the Benedictines faced the problem of how to ensure that they stay on schedule. They found their answer in the invention of the mechanical clock circa 1300 – an automated machine that ran by a device called an escapement, a mechanism that "regularly interrupted the force of a falling weight," controlling the release of energy and the movement of the gears.[6] The mechanical clock was a godsend, allowing the brethren to standardize the length of hours and stay on time and on schedule during their daily activities.

The invention of the mechanical clock was so alluring that word of the machine quickly spread beyond the cloisters and into the public domain, with a giant clock looming over every town square and becoming the centerpiece of daily life and, soon thereafter, the coordinator and orchestrator of a proto-industrial economy in urban communes across late Medieval Europe. Of course, the Benedictine monks never intended to have their time-controlling invention become the overseer of commercial life and a new temporal orientation that would come to be thought of as efficiency. The mechanical clock was an ideal invention to manage, monitor, and sequester a proto-industrial economy, where relentless efficiency would come to replace the more relaxed temporal orientation of small urban communes and an agricultural economy, where time was measured by sunrise, high noon, sunset, and the changing seasons of the year. Calibrating every moment of time quickly became a fetish in the industrial era. The minute hand was soon introduced, and, shortly thereafter, the second hand. With the rise of the Industrial Age and market capitalism, "Time is money" became the new overarching mantra of social, civic, and commercial life. By the 1790s, timepieces, once a spectacle and luxury, had become affordable necessities in every home, and workers were even beginning to wear pocket watches.

In Jonathan Swift's *Gulliver's Travels*, the wise men of Lilliput informed the emperor that the alien giant they had shackled reached into his pocket and took out a shiny object that made a humming noise like a miniature windmill, and "put this engine" to their ears. They thought that the machine was "either some unknown animal, or the god that he worships; but we are more inclined to the latter opinion, because he assured us . . . that he seldom did anything without consulting it."[7]

The clock and pocket watch were steadily reorienting the public away from nature's time and toward the steady beat of mechanical time on the factory floor where production systems required unerring attention in the synchronization of industrial activity. Time would come to be thought of as standard measurable units operating in a parallel universe far removed from the rhythms of an animated Earth. Although little considered and less thought of, efficiency took our species into an alternative virtual world, completely detached from the temporal rhythms that marked the daily rotation of the planet and the changing seasons, as the Earth made its yearly passage around the sun.

But there is more to this story. In the Medieval era, time on earth was of little account. To be "in the world" but not "of the world" defined the period. The earthly plane was viewed as a temporary holding pen as the Christian faithful anxiously awaited their ascent to heaven and everlasting life.

We only need to take a look at the paintings and tapestries of the period which depict the Creation as a flat plane, with human beings floating upwards with outstretched hands and eyes glued to heaven above. While beautiful, these paintings appear dreamlike and childish and lack depth. They don't appear realistic.

That all changed when the artist Filippo Brunelleschi tore up the Church's script in 1415 by being the first to use linear perspective in art to depict the baptistery in Florence from the front gate of the cathedral while it was still under construction. The linear perspective projected the illusion of three-dimensional depth by use of a "vanishing point" to which all lines converge, at eye level, on the horizon.[8] Brunelleschi's stroke of genius touched off a mind-turning perception in how our species would come to think of both temporality and spatiality, marking one of the great changes in the way we view the world. Michelangelo, Raphael, Donatello, and Leonardo da Vinci soon put brush to canvas, painting their own masterpieces depicting perspective. The introduction of perspective in art on the heels of the mechanical clock literally changed the way successive generations would come to experience time and space and existence on Earth.

From now on they would be looking at paintings that repositioned the eye to gaze out from behind a window at an empty horizon ready to be captured, sequestered, objectified, and transformed to create a second Eden on Earth. Every person becomes both a voyeur and overseer of everything in his or her gaze. Perspective in art prepared generations that followed to think of the world from the point of view of a detached and objective observer.

The artists of the Renaissance, who brought perspective to the medium, often interacted with architects and mathematicians of the day, who used the principles of perspective to advance their own disciplines, while sharing the insights of geometry and the science of the day with artists. These technical changes in how successive generations would come to perceive existence ushered Baconian science onto center stage and laid the groundwork for the Age of Enlightenment in the eighteenth century and, thereafter, the Age of Progress in the nineteenth and twentieth centuries.

Generations of our species would come to see the world around them as objects to capture and utilize – in other words, passive matter ready to be transformed and devoured. And utilitarianism would come to replace, at least in part, an earlier religiosity. To see the world as a detached observer is to remove oneself from one's surroundings and assume the role of an architect and entrepreneur, right down to our most trivial interactions with the planet. Even in the most mundane aspects of our day-to-day lives, we each become an overseer, and the natural world around us is viewed as a utility for consumption.

The notion of molding time to the requisites of efficiency and utilitarianism has become so interwoven that for all intents and purposes they appear tautological. Efficiency is the marker of the modern notion of temporality and has enjoyed an unquestioned status as the paramount virtue of contemporary life. Its dominance as a temporal value has gone mostly unchallenged, as if to suggest that it's the primary temporal medium underlying nature itself.

We have come to view efficiency as a force of nature instead of a human invention. Breaking it down to its bare essentials, when we think of efficiency, we envision the extracting, sequestering, commodifying, and consuming of ever-greater volumes of the Earth's guiding spheres – the hydrosphere, lithosphere, atmosphere, and biosphere – at ever-faster speeds and in ever-shorter time intervals to increase the exclusive opulence of our own species. Doesn't every other species do the same? Hardly . . . in nature, the temporal value is not efficiency but adaptivity. Every creature and even our own species, at least until the dawn of civilization, continually adapted its own biological clocks – ultradian, circadian, seasonal, and circannual rhythms – to the temporal orientation of the planet.

Efficiency, then, is a purely human invention – a temporal value that does not exist in nature. In nature, adaptivity is the universal driving temporal value – it's built into the genetic makeup of every living creature. Likewise, "productivity," efficiency's twin, is nowhere to be found in nature. Rather, "regenerativity" drives natural systems and all of the life forces that abide. And nature is not made up of passive resources and objects to commodify and consume but a rich reservoir of dynamic processes, patterns, and flows continually interacting in a self-organizing theater of engagement that makes up a living Earth.

While this other experience of nature as animated, interactive, and ever self-evolving over time, rather than passively existing in an atemporal spatial vacuum is the way our species perceived our existence from time immemorial, the great departure that took us away from nature's calling began inauspiciously with the rise of the first urban hydraulic civilizations some six thousand years ago. That saga is now upending, as a warmer climate is unshackling the climate's hydrosphere as it seeks a new course.

To be fair, there have been spurts of recognition of how the real world of time and space actually operates – most recently, during the Romantic Period's counter revolution to the Age of Reason and the Enlightenment. But it wasn't until the early decades of the twentieth century that physicists began to realize that their earlier suppositions about the physicality of atoms as solid matter taking up a fixed space had been "misplaced." Scientists came to realize that an atom is not a "thing" in the material sense but a set of relationships operating at a certain rhythm, and that being so, "within a given instant of time the atom doesn't possess those qualities at all." As the physicist Fritjof Capra explains:

> At the subatomic level, the solid material objects of classical physics dissolve into wave-like patterns of . . . probabilities of interconnections. Quantum theory forces us to see the universe not as a collection of physical objects, but rather as a complicated web of relations between the various parts of a unified whole.[9]

The oft-held practice of separating structure from function peeled away with the birth of the new physics. Scientists began to realize that it's impossible to separate what something is from what it does. Everything is pure activity, and nothing is static. Things do not exist in isolation, but through time.

It was after World War II that Norbert Wiener, the father of cybernetics, and his contemporary Ludwig von Bertalanffy, the father of general systems theory, stepped in and put theory to practice. Each came to believe that humanity's long-held assumptions about time, space, and the nature of existence were misconceived. In 1952, von Bertalanffy wrote, "What are called structures are slow processes of long duration, functions are quick processes of short duration." In 1954, Wiener took a similar approach to sizing up our species and, for that matter, the life of every other species inhabiting the Earth. He wrote of human life that:

> It is the pattern maintained by this homeostasis, which is the touchstone of our personal identity. Our tissues change as we live: the food we eat and the air we breathe become flesh of our flesh and bone of our bone, and the momentary elements of our flesh and bone pass out of our body every day with our excreta. *We are but whirlpools in a river of ever-flowing water.* We are not stuff that abides, but patterns that perpetuate themselves.[10]

Their musings owe much to the thinking of the philosopher Alfred North Whitehead. Whitehead is regarded as one of the great mathematicians of the twentieth century. He coauthored with Bertrand Russell the *Principia Mathematica*, a three-volume series on the foundations of mathematics that became math's go-to reference in the twentieth century. In later life, he shifted his focus to philosophy and physics. His principal work, *Process and Reality*, published in 1929, went on to influence many of the leading thinkers in science and philosophy over the course of the century.

Whitehead pilloried Isaac Newton's description of matter and motion devoid of the passage of time:

> which presupposes the ultimate fact of an irreducible brute matter, or material, spread throughout space in a flux of configurations. In itself such a material is senseless, valueless, purposeless. It just does what it does do, following a fixed routine imposed by external relations which do not spring from the nature of its being.[11]

Whitehead took exception to Newton's description of existence as made up of "durationless" instants "without reference to any other instant," arguing that "velocity at an instant" and "momentum at an instant" was simply absurd.[12] Whitehead argued that the idea of isolated matter having "the property of simple location in space and time" left "Nature still without meaning or value."[13]

What troubled Whitehead was that the prevailing worldview of Nature in the scientific community "omits any discrimination of the fundamental activities within Nature."[14] The historian and philosopher Robin G. Collingwood of Oxford University observed that relationships and rhythms only exist in "a tract of time long enough for the rhythm of the movement to establish itself."[15] To cite the obvious – for example, a note of music is nothing without the notes that precede and follow it.

Whitehead summed up the new view of physics:

> The older point of view enables us to abstract from change and to conceive of the full reality of Nature at an instant, in abstraction from any temporal duration and characterized as to its interrelations solely by the instantaneous distribution of matter in space . . . For the modern view process, activity, and change are the matter of fact. At an instant there is nothing. Each instant is only a way of grouping matters of fact. Thus, since there are no instants, conceived as simple primary entities, there is no Nature at an instant.[16]

Assume our scientists are right in understanding that all of nature is ephemeral and what we experience are ever-evolving patterns, processes, and flows prompted by the hydrosphere and impacting the lithosphere, atmosphere, and biosphere on a self-organizing planet, replete with the Butterfly Effect and externalities that accompany every movement, continually rearranging what we've come to think of as existence. If so, we have to rethink the very idea of nature as passive, atemporal objects and structures that can be stripped from their surroundings and captured, sequestered, propertized, and commodified at will. To put it bluntly, this simply isn't the way the planet operates and functions.

The point is John Locke and later Adam Smith and the acolytes that followed in their footsteps in the framing of capitalist theory and practice have egregiously misunderstood the planet we live on. More importantly, their philosophical missteps have resulted in the pillaging of an animated and alive planet, taking our species and our fellow creatures into the sixth extinction of life on Earth. But now it's the arts, following in the footsteps of the new understandings in physics and biology, that are helping our collective humanity to reimagine what life is all about on Planet Aqua.

Once again, as was the case in the Italian Renaissance, the arts are breaking the mold and forcing a rethink in the way we conceive of time and space and, by extension, our relationship to the natural world, our notions of selfhood, our approach to science and technology, and even the way we educate our youth, extract economic life from the Earth's great spheres and conceive of governance.

Ephemeral art is as old as our species and draws on 200,000-plus years of our existence on Earth. Our early forager-hunter ancestors lived a life of pure ephemerality. As mentioned earlier, their very persona was of an animist nature. They experienced a world inhabited by a theater of fluid interacting forces, patterns, and flows imbued with agency and intimately affecting every aspect of their existence. The raging rivers and streams, the dark cool forests alive and swaying in the wind, the imposing snow-crested mountain ranges peering down from above, the powerful jettisoning winds sweeping across the horizon, and even the fresh footprints of fellow creatures left for a passing moment were perceived as spirits in the milieu in which our species dwelled – undifferentiated. These other spirits and demons even accompanied our ancestors to the netherworld – a place absent of corporality.

In our ancient ancestors' world, the visual, auditory, olfactory, taste, and haptic senses were continually drawing from a rich sensual environment that was continually reorganizing and evolving with each passing moment and in the process taking them along for the journey. The world of forager-hunters was ephemeral and forever improvisational. The concept of space as place, a fixed unchanging milieu, would have been inconceivable to forager-hunters living a nomadic existence. If there was any sense of constancy in this highly animated, ephemeral cosmology, it was the passing of the seasons and the marking of the winter and summer solstices. Mircea Eliade referred to the ethereal nature of the yearly cycle of birth, life, death, and rebirth of existence celebrated by our animist forebears as the "eternal return."

Just as their migratory life was wedded to the birth, life, death, and rebirth of the seasonal cycles, so too did they come to understand their own passages in life. Upon death, one's spirit remains in limbo in the underworld, only eventually to find its way into other forms of life, be it human or other creatures, or even embedded in the inanimate world. The nineteenth-century anthropologist Sir Edward Tylor was the first to categorize such societies as animist cultures. In short, our ancestors' world was ephemeral and what we would expect of a nomadic forager/hunter culture.

The changeover from foraging and hunting to agriculture and pastoralization and thereafter the rise of urban hydraulic civilization six thousand years ago led to a sedentary existence and with it, at least in the Western world, to a bifurcation of consciousness, pitting "place" above "movement," and "being" against "becoming" and first articulated by Plato and the Greek philosophers of yore.

How Plato Took Our Species Down the Wrong Path

Plato is the founding father of deductive reasoning, the bedrock of philosophy. The Greek philosopher divided existence into two realms, the first occupied by what he called "Ideas or Forms" which are the nonphysical essences of the cosmos, and the second the realm of matter, objects, and things. The second world, the physical one we live in, says Plato, is but a pale imitation of the first world of Forms which is "immaterial, non-spatial, and atemporal" and whose essence is pure being.[17]

According to Plato, every single object we experience and interact with in the day-to-day world is an imperfect imitation of these universal Forms. For example, individual human beings, mountain ranges, pieces of art, are all pale copies of the ideas they represent. A person's experience of love is a pale representation of the Form love. The drawing of a triangle is a pale imitation of the concept of a triangle. Each of the above examples is temporal in nature, while Plato argues that Forms do not exist in either time or space and are invisible.

No one has ever seen a perfect triangle or a perfect representation of beauty, or for that matter a perfect dog. Still, they each exist as archetypes that can only be imagined as concepts in one's mind. No human being can experience or observe these Forms but can only have an idea in their head about them which they can then imitate. We can "think" of what a perfect triangle might be, but we can never "experience" it. Plato would argue that knowing truth can only be experienced by pure thought and deductive reasoning alone and not by sensory experience.

Plato introduced into Western philosophy the concept of the mind–body split – the separation of the cognitive and physical worlds that has come to shape the way that generations of scholars, and especially scientists, have conducted their ontological inquiries. We are all familiar with the oft-heard quip "try not to be so emotional . . . be more rational. Trust reason over experience." In philosophy, rationalism is the methodology by which we use reason to secure knowledge "in which the criterion of truth is not sensory but intellectual and deductive."[18]

For Plato then, reality is not the world we experience through our senses but the very abstract and non-material world of ideas or Forms, which always remain beyond our grasp. Perhaps the closest one can get to experiencing pure reason is mathematics, which allows one to glimpse universal truths without relying on sensory experience. Plato was enamored with geometry – at the time the most sophisticated genre of mathematics – and viewed it as a window to pure knowledge. He even had the phrase "let no one ignorant of geometry enter" inscribed over the door of his academy.

The great philosophical juggernaut, in all of its manifestations, with a detached rational humanity overseeing and sequestering an atemporal passive nature, devoid of any sense of agency, has accompanied an urban culture increasingly detached from the natural world all the way to the current era.

To put the record straight, it was Isaac Newton who hammered the last nail into the coffin of a temporal world. Newton uncovered the mathematical formula for describing gravitation. He posited that a single law could describe why the planets move in a certain fashion and why an apple falls from the tree in a particular way. Newton argued that the phenomena of nature "may all depend upon certain forces by which the particles of bodies, by some causes hitherto unknown, are either mutually impelled toward each other, and cohere in regular figures, or are repelled and recede from each other."[19] According to Newton's three laws: a body at rest remains at rest, and a body in motion remains in motion in a straight line unless acted upon by an external force; the acceleration of a body is directly proportional to the applied force and in the direction of the straight line in which the force acts; and, for every force, there is an equal and opposite force in reaction. Newton's three laws deal with how all the forces in the cosmos interact and settle back into "equilibrium."

Newton's universe of matter and motion was orderly and calculable and made no room for the temporal world of spontaneity or unpredictability. It was a world of quantities without qualities. Newton mathematicized the Age of the Enlightenment, and mathematics, in turn, provided the scaffolding for the ensuing Age of Progress. Most importantly, Newton's three laws of matter in motion are absent time's arrow. In Newton's universe, all processes are time reversible. Yet, in the real world of nature and, by extension, the economy, no event is time reversible. By embracing Newton's atemporal schema as a tool to model economic activity, generations of economists would be led astray, further distancing themselves from the real world.

A fast-warming climate and a rewilding hydrological cycle is now the new normal. Suddenly the earth's great spheres – the hydrosphere, lithosphere, atmosphere, and biosphere – thought of as atemporal domains to be extracted, sequestered, propertized, commodified, and consumed have come alive with agency, even though that agency was always there but unrecognized during the long stretch of a mild climate and seemingly docile nature.

And it's the arts that are leading the way, rejecting the Renaissance legacy of objective detachment and the capture of nature as fixed objects without temporal agency. The ephemeral arts are reintroducing our species to an animated Earth full of temporal agency, with surprises at every turn. A new generation of artists is breathing life back onto the Earth by introducing us to an unpredictable and ever-evolving planetary milieu of interacting forces, processes, and patterns emerging moment to moment and then quickly submerged, making way for new twists and turns in what has always been an animated planet.

This new generation of ephemeral artists is giving us a peek into a planet alive and full of agency, attempting to adjust to a changing climate. The reintroduction of ephemeral arts after a long period of near dormancy helps us to understand the transient nature of existence and the importance of every precious moment in each of our sojourns, along with the knowledge that our individual experiences live on and affect everything that is yet to come.

Ephemeral art has been with us since our ancient ancestors stepped out of the forests in the Rift Valley of Africa some 200,000 to 300,000 years ago and onto the open savannas, and from there trekked across the vast oceans, settling the continents. Ephemeral art only began to fade – but never really died – with the rise of urban civilization and our species' increasing separation and detachment from the natural world in the millennia before Christ. Today, over half of the human family is tucked away in artificial urban enclaves and, for the most part, locked onto virtual screens far removed from what might be identified as nature. Still, the ancient ephemeral arts have remained, more so in developing countries and among religious sects and indigenous communities, and less so in highly urbanized industrial nations, increasingly detached from a world of nature.

The acting out of elaborate ephemeral rituals steeped in the arts and generally paying homage to the spirits, goddesses, and gods of nature who have blessed them with the Earth's generosity continue to take place in many countries. Often the body is transformed into an ephemeral work of art adorned with feathers and fur with the face and skin colored in elaborate designs, using natural pigments taken from crushed stones and soils in shades of yellows, browns, oranges, reds, and blues.

Hindu legend, for example, pays homage to Multani mitti clay, which is deemed to have magical properties, because from the earth "all creation is fashioned, and to Earth it eventually returns."[20] It is said that Multani mitti is "the body of the mother goddess, the sustainer of life" and is a substance that "is easily procured and sculptured and, when purified by fire or other sacred elements it becomes an appropriate vessel for facilitating Darshan with the gods."

In eastern India, an ancient ritual is reenacted each year, with potters applying clay to elaborate straw-and-stick structures to be carried in processions to the local river where they dissolve back into the current, allowing the deity to return to the heavens as the ephemeral sculpture fashioned of the Earth is returned to the Earth, having served its purpose. Similar ancient ephemeral rituals take place all over the world. Body paintings, sculptures fashioned from the Earth, and dance rituals are all parts of the oldest ephemeral works of art and performance, and each reflects the transitory nature of life, whose every moment dissolves back into the pool of nature living on in other guises, having left an impression on every moment yet to come.

It's instructive to understand the distinctions between the traditional arts and the ephemeral arts. The traditional arts include architecture, dance, sculpture, music, painting, poetry, literature, theater, narrative, film, and photography. While some of the traditional arts are ephemeral – for example, dance, theater, and the recitation of poetry – the moment they are photographed, filmed, written down, and archived, they become frozen in time and space.

The ephemeral arts, by contrast, are alive in time and space and, while retained in memory, are not archived. Rather, they come and go, leaving a signature that ripples out with agency, affecting everything that comes thereafter. Regardless of how significant their contribution, they are fleeting experiences in time. The term "ephemeral" is derivative of the Greek word "ephemeros," which means "short-lived." Sand art and ice sculptures would fall under the category of ephemeral arts. Ephemeral art is intended to be immediate, dissipating, and not preserved. This form of art celebrates the temporality of existence. By contrast, conventional arts, including painting, sculpture, and photography capture an image in space and sequester it as an atemporal object giving it the veneer of permanence.

The Eastern religions and philosophies were far more attuned to the ephemerality of existence in their artistic expressions than the West. For example, their religious rituals and philosophical practices often included creating beautiful and intricate mandalas, composed of colored sands, which upon completion were scattered to the winds, reflecting the belief that existence itself is fleeting and ephemeral.

The German philosopher of the early twentieth century, Walter Benjamin, reasoned that the shift to ephemerality in works of art in recent times reflects the speeding-up of life in a fast-changing global milieu. By the 1960s and 1970s, a new wave of ephemeral art began to emerge, as a younger generation became increasingly aware of the deleterious changes taking place in the natural environment. The idea of a relatively predictable nature made up of passive objects long taken for granted conjured up an atemporal world of pure being, while the raucous rewilding of the Earth's hydrological cycle brought on by global warming made every moment seem unique. A radically morphing hydrosphere has pierced the consciousness of our collective humanity, forcing each and every one of us to experience nature as unpredictable and continually evolving.

The modern ephemeral wave in the arts has roots in African-American culture. The Jazz Age in the 1920s brought ephemerality into popular culture and changed the very way that music is approached. Pick-up groups would join together without a musical roadmap and improvise moment-to-moment, making up the music spontaneously and transforming each musical encounter into a unique and unrepeatable ephemeral experience. Improv comedy took off in the 1960s in Chicago with the Second City Players, many of whom went on to pioneer the still-popular TV show, *Saturday Night Live*. Improv comedians generally would begin with a loose storyline and let go with a kind of flow of consciousness, typical of the consciousness-raising therapeutic encounters popular at the time in women's consciousness-raising sessions. Improv comedians would often lay bare the ironies, mishaps, and Catch-22s of life that sting, and bring a laugh of recognition from audiences who relate to similar experiences in their own lives.

Around the same time, Julian Beck and Judith Malina brought the concept of living theater to the New York stage and, soon thereafter, the world stage. Each performance centered around a particular theme or storyline, and the audience was sometimes invited onto the stage to improvise with the actors in a kind of theatrical jazz.

Dance improv became a popular genre in the 1960s. Letting go of the strict formatting found in ballet and in other forms of the dance and allowing the performer to improvise movements cued to the feelings of the moment and the context of the environment became a kind of therapeutic exercise and an ephemeral art form. Martha Graham, Doris Humphrey, Paul Taylor, and Merce Cunningham, among others, pioneered this new animated dance jazz. Yvonne Rainer's Grand Union Dance Group in the 1970s took ephemeral dance a step further in performing improv that was purely spontaneous and not rehearsed beforehand.

But it wasn't until the late 1960s and the awakening of the modern environmental movement with Rachel Carson's publication of the book *Silent Spring* and an emerging environmental crisis, followed by the first Earth Day celebration in 1970 that a small coterie of artists left the art museums and art galleries altogether and began practicing a new genre of ephemeral art called "land art." Claude Monet, the great impressionist artist of the nineteenth century, once regaled, "my garden is the most beautiful masterpiece" but, of course, all gardens are ephemeral works of art, continually morphing moment-to-moment across the seasons, allowing a gardener to enjoy his or her participation as a steward of nature's lifeworld in an act of co-creation.

Land art is not a new phenomenon. It has been practiced as long as our species has inhabited the planet. But it has taken on new meaning in the throes of the sixth extinction of life on Earth. Land art is a celebration of the awe and wonder of existence – the relentless self-organizing evolution of an animated Earth that has taken millions of species – our own included – on a journey across new horizons.

The new generation of land artists pay homage to the ephemerality of a living Earth. To appreciate the significance of ephemeral land art and how it alters our notions of time and space, consider two very disparate approaches to sculpture, each affecting a very different emotional response to the awe of existence. The first sculpture and, arguably, the most recognizable sculpture in the world is chiseled out of the rock atop Mount Rushmore in South Dakota in the United States. The sculpture depicts the heads of four of the most beloved presidents of the United States: George Washington, Thomas Jefferson, Abraham Lincoln, and Theodore Roosevelt. Each sculptured head is sixty feet tall and looms over the landscape and can be viewed from miles away. The sculptures represent the country's "birth, growth, development, and preservation" and draw more than two million visitors each year.[21]

The chiseling of the sculptures that began in 1927 and was completed in 1941 was not without controversy. The indigenous Sioux Nation argued that the sculptures lie on sacred Sioux land that represents its ancestral deities personified as the six directions: "North, South, East, West, Above (sky), and Below (Earth)" and have referred to the sculptures as the "shrine of hypocrisy."[22]

The "four grandfathers" of the United States of America chiseled into the granite peak of Mt. Rushmore comes with an even deeper yet left unsaid meaning – the idea of "man's commanding dominance over nature." The underlying but unconsciously felt theme is man's conquest of time and the sequestering of space. No doubt, many a spectator leaves the site with the thought that the giant sculptured granite faces, like the ancient pyramids of Egypt, will be here forever – a paean to man's immortality and the subjugation of temporality – eternal being triumphing over earthly becoming.

Land artists take a very different approach, seeking to interact with the temporality of nature. Their world is one of continuous becomings. They perceive the Earth's four great spheres – the hydrosphere, lithosphere, atmosphere, and biosphere – as animated with agency all intertwined in an earthly temporal cauldron continually evolving into new patterns that live on and spread out, affecting everything else regardless of how slim the externality.

Andy Goldsworthy is a British sculptor of a different ilk. Among the pioneers in the new artistic realm of ephemeral land art, his sculptures use snow, ice, fallen leaves, flowers, pinecones, and stones as feedstock from which to fashion elaborate and beautiful works of art embedded in the natural environment that play with light and shadows in ways that are alluring but also temporary, only to be whisked away by the wind, water, heat, or the slow degradation into the Earth's pedosphere. His sculptures are hidden in the forests or on the edge of floodplains and wetlands waiting to be discovered. He is best known as the father of "modern rock balancing . . ." rocks arranged in elaborate towers that seem to defy gravity. Passers-by are struck by the sudden discovery of these installations, piquing their curiosity and even their sense of awe and wonderment. These chance encounters often leave a mark on people's memories and live on in the stories they share with others.

Most interesting, the ephemeral land art installations escape commodification. They are to be experienced for a moment in time and appreciated but cannot be sold and exchanged. While the installations degrade over time, the atoms that make up the molecules that take up residence in these ephemeral sculptures will eventually move on and take up residence somewhere else and in some other temporal form . . . in this sense, they never die.

By denigrating lived experience as a pale imitation of pure thought, Plato, in effect, threw to the wayside the physicality of life on an animated Earth charged with agency and brimming over with novelty and emergence. His detached rationalized world of pure thought and perfection unencumbered by lived experience and the passage of time shares a close affinity with the notion of heaven. "Eternity," after all, is a timeless realm populated by spirits devoid of physical senses and existing in a state of perfection without change.

At least Plato was not shy about identifying with a detached rationality as the essence of the eternal and devaluing the physicality of ephemeral existence as of little significance in the bigger picture of things. He even disparaged physical beauty for its transience, arguing that only "the idea of the beautiful" is eternal. To be sure, prizing the ephemerality of life didn't altogether disappear, but remember that for nearly seventeen centuries since the crucifixion of Christ and his ascension into heaven, the Christian faithful remained huddled and steadfast in wait, praying to be lifted up from a fallen physical world of toil and turmoil and welcomed into the heavenly realm of eternal life.

Toward the end of the eighteenth century, the Romantic Period was stirring and its enthusiasts had enough with all the footnotes to Plato, making the often-heard remark, "the beautiful will never be seen twice." By the late 1960s, the tables had turned and ephemerality was experiencing its own second coming – a more sophisticated neo-animism attached to the new findings in physics, chemistry, biology, and the ecological sciences. Again, Walter Benjamin glimpsed the spirit of the times to come, arguing that in modernity, it's the transitory nature of nature that is eternalized.[23]

Why devote so much ink to the resurgence of ephemeral art? Because it is preparing our species for living on a super-charged climate-changing planet that's going to require us to surf the waves of an unpredictable hydrosphere, taking us into unexplored new worlds. The sedentary life of the past 10,000 years, with its mostly predictable mild climate, is now passing. It was always a short interlude in our species' existence on Earth. Learning to live and thrive as neo-nomads in a world of constant and harrowing change touched off by the rewilding of the waters requires a reset on how we experience time and attachment to place. And this means giving up the sophomoric notion that nature is made up of passive objects without agency that can be corralled and consumed without negative externalities that ripple out and come back to haunt us. This way of thinking is at best naive and, at worst, endangering our ability to endure and flourish.

A new more mature neo-animism is being born, giving our species a lifeboat to the future as we prepare ourselves for a more nomadic existence tempered with brief periods of sedentary life – a future quite different from what we have known during the short period of a predictable mild climate, where attachment to place and a more sedentary existence ruled. If there is any hope to hold on to, it's that our biologists and anthropologists tell us that our species is biologically suited to live as our ancient nomadic ancestors did over eons of history. Our biology is our strong suit that will serve us well and even allow us to persevere on a highly animated planet that's rewilding in new ways. Reorienting ourselves to the temporality of existence is the first step in preparing for a more ephemeral and nomadic life.

CHAPTER 9

Rethinking Attachment to Place

Where We've Come from and Where We're Heading

Our biological anthropologists still aren't exactly sure when *Homo sapiens* arrived on the scene. For a long time, the prevailing wisdom was that we emerged around 200,000 years ago in the Rift Valley of Africa. More recently, new fossil finds and genetic profiling suggest that our ancient ancestors might date as far back as 300,000 years. What they all do agree upon is that from the very beginning we were a highly nomadic creature and, for the most part, chasing the seasons and climate with short periods of respite, but what could hardly be called sedentary life – that would have to come with the melting of the last Ice Age and the emergence of a mild and mostly predictable weather and climate of the past 11,700 years. Here and there, small bands of our ancestors settled down and instead of tracking with the seasons and the shifting vegetative cycles and animal migrations they learned to domesticate wild plants and corral wild animals and pasture them. Instead of adapting to nature, they began the big turnaround by adapting nature to our species.

The first agricultural settlements abutted ocean coastlines or nearby lakes, rivers, and wetlands, to ensure an abundance of marine life and a dependable storehouse of water for human consumption and primitive irrigation. But to believe that sedentary life began to overtake migratory life would be a mistake. That's never been the case. If we step back and look at the long stretch of *Homo sapiens*' footprint on Earth, it's clear that while sedentary life, first in villages and later in cities, and now in megalopolises, has carried the bulk of our population, it's also true that great civilizations have risen and fallen, come and gone.

For the most part, the sedentary infrastructures were the victims of entropic dissipation from the degradation of the soils, the ravishing of the forests, and the depletion of the waters. But they were also skewered by dramatic changes in the climate throughout our species' historical journey. Invasions by marauding bands and colonizing forces have also resulted in the scattering and migration of our species over time. More often than not, it's been the morphing of the Earth's great spheres – the hydrosphere, lithosphere, atmosphere, and biosphere – that have been the ultimate arbiters of our back and forth between short sedentary stays and longer nomadic periods.

Consider that in 1776 our species numbered less than 780 million people. Today we number eight billion people, brought on by a fossil fuel-based water–energy–food nexus industrial civilization that is dramatically altering the Earth's climate dynamic, already throwing hundreds of millions of people into the throes of mass migration and resettlement. By the end of the current century several billion human beings might be migrating in search of safer climate havens. Even before the dramatic warming of the climate began to affect everyday life, the population bell curve was beginning to peak. The Wittgenstein Center for Demography and Global Human Capital in Vienna projects that by the year 2080 the global population will peak at around 9.4 billion people and begin a sharp decline over the span of the next century.[1]

The tilt from sedentary to nomadic life and the hardships that accompany mass migration on a climate-warming planet will inevitably lead to a precipitous decline in the human population. That process has already begun.[2] By the twenty-fourth century, it's likely our population will have declined to levels more in line with our pre-industrial population, as we fully transition into the new green energies brought on by the sun, the wind, and the waters, which are the primal energies that govern this planet.

Stepping into Neo-Nomadism

The world we have known during the whole of the so-called "Age of Progress" is a fleeting affair and is quickly coming to an end and much faster than we think. Where then does hope reside? From our emergence in the African savannahs, modern humans began the long trek across the planet. Our ancestors first went east, crossing all of Asia and into Indonesia 80,000 to 60,000 years ago and from there migrated to Australia. Around 45,000 years ago, our ancient brethren traversed the Mediterranean and traveled along the Danube into Europe, where they lived side-by-side with Neanderthals with whom they sometimes mated and whose genes are still found in our genetic makeup.[3] It was only 15,000 years ago that *Homo sapiens* crossed from Asia into the Americas via the Bering Strait and drifted down through North, Central, and South America.

The early anthropologists put great store on our ancestors being drawn in their migration to lush grasslands, where there was substantial prey in open fields for the taking in the hunt. Though not altogether misleading, our forebears were less hunters and more foragers because wild animals were scarce, while vegetation was readily available. Patrick Manning, a professor of history and African studies at Northeastern University in the United States and author of *Migration in World History,* points out that the great migration and settling of populations across continents in the early Neolithic era favored homesteading near water edges. He writes:

> as human communities grew and spread, they were confronted repeatedly by a choice: concentrate at water's edge and range across open grasslands . . . studies of human evolution have long tended to emphasize hunting and the grasslands . . . I want to emphasize the continuing importance of rivers, lakes, and oceans among early *Homo sapiens*. Gatherers found a rich variety of plant and animal life along the seashore, along rivers, and lakesides.[4]

The question that has puzzled anthropologists is exactly how did our ancient ancestors cross the great oceans and settle new landmasses and entire continents? While the short hop from Asia to the Americas over the Bering Strait is only 53 miles wide and relatively shallow in parts, it still leaves unanswered how our ancestors managed sea voyages across vast stretches of the great oceans, for example from the Indonesian archipelago to Australia and the Pacific Islands, many of which were settled thousands of years ago.[5]

The first great navigation of the oceans, at least on the Pacific side, was launched from Polynesia, a subregion of Oceania comprising more than 1,000 islands stretching across the Pacific Ocean. "Polynesian wayfinding" dates far back into history.[6] Their peoples became the first ocean travelers, colonizing islands and engaging in complex trade. Particularly fascinating is their sophisticated navigational skills. Relying on memory, they carefully archived certain sea birds, who would fly out onto the ocean to fish and return to land in the night hours, allowing the navigators to follow suit. They also navigated by the stars. These savvy navigators even used waves and swell formations to navigate the oceans, finding their way from one island to another. Navigators who lived within a group of islands would learn the effect various islands had on the swell shape, direction, and motion, and would have been able to correct their route accordingly.[7]

Human migration through history has taken on various personas: fleeing oppression and evading wars, colonizing other populations, seeking out virgin lands to homestead, creating new communities for religious practices, the search for new places to work, migrating from barren lands and food deficits, fleeing from religious, ethnic, racial, and political persecutions, escaping from climate impacts including floods, droughts, earthquakes, and volcanic eruptions. Poverty and the hope of greater opportunities have been a mainstay of mass migrations. Between 1820 and 2013, 79.5 million people – mostly poor, hungry, and destitute – migrated to the United States and retained residence status and citizenship.[8] The movement of people also included the forced transporting of millions of slaves from their homelands in Africa to regions throughout the Americas. Mass migrations were also spurred by the desire to affiliate with a different way of life and culture. There is also the dynamic of internal migration. In the U.S., six million African Americans journeyed from the southern states to northern states between 1910 and 1970 in what was called "The Great Migration," fleeing from white supremacy and Jim Crow laws.

In recent times, the opportunities of seasonal labor have drawn migrants from poorer nations to their richer neighbors where they work and also send remittances back to their families in their native countries. For example, a large number of citizens of the Philippines migrate to the United Arab Emirates for extended periods of time to work and then return to their host country. There are also those who are migrating to seek new adventures or more cosmopolitan lifestyles or solitude.

While reasons for most migrations are not clear cut, we need to distinguish between the various degrees of migration that separate sedentary life from nomadic life. Even long-distance commuting for work or traveling between two regions during summer and winter seasons on holiday blurs the line between what is sedentary and what is nomadic. How, then, do we draw a line between our nomadic and sedentary selves, for example, between visiting distant friends and relatives for extended periods of time, and returning? Or consider this. An eighteen-year-old adult in the United States is likely to move 9.1 times during his or her lifetime, often to regions far removed, according to the U.S. Census Bureau.[9] It's also interesting to note that 281 million people or 3.6% of the world's population lived outside their country of citizenship in 2020.[10] If there is any doubt as to our nomadic drive, consider that in 1960 only one-third of countries allowed dual citizenship but, by 2019, 75% of countries allowed dual citizenship, and that number is increasing yearly.[11]

Homo sapiens are by nature nomadic, with only sedentary interludes. While settling down is coveted, the itch to explore is wired into our very physiology. The numbers speak for themselves. In one typical recent year – 2017 – before the COVID-19 pandemic, 1.3 billion people were counted as international tourists.[12] In 2019, travel and tourism accounted for one out of every four new jobs across the world and 10.3% of all employment . . . that's 330 million jobs in total. The industry represented 9.6 trillion dollars of global GDP, or around ten percent of global GDP, and visitor spending amounted to $1.8 trillion or 6.8% of total exports.[13]

Nomadism is baked into our physiology and has been critical to our flourishing and survival throughout history. Why is this so important to acknowledge? Because we are facing the greatest mass migration in human history in the throes of a global warming climate and the sixth extinction of life on the planet. Several billion of our human family will likely be fleeing climate hotspots in the future. The search for new safe havens and a place to flourish will be the overriding concern for eons to come. Those who come after us will live a more nomadic lifestyle, with shorter interludes of sedentary life in between, as the planet's hydrosphere rewilds.

When asked how we survive and hopefully thrive in new ways amid a mercurial and unpredictable change in the Earth's climate, we will need to realize that we are in fact among the most adaptive species on Earth, with the exception, perhaps, of viruses. The Smithsonian Institution and New York University researchers conducted an unusual study, which asked how our diminutive species survived and prospered over its short period of time on Earth.[14] As mentioned, the old story that generations have been told is that around 11,000 years ago the last Ice Age receded, and a new temperate climate took hold. When that happened, the nomadic way of life of foraging and hunting gave way to a sedentary way of life of farming and pastoralization, followed by the great hydraulic civilizations and urban life, and finally the Industrial Age and the near total urbanization of our species, anchored down in a sedentary existence. But when the researchers looked into the geological record, they found, much to their surprise, that when our hominid ancestors emerged some 800,000 years ago, and much later our Neanderthal brethren, and eventually *Homo sapiens,* they lived during one of the most violent changes in weather patterns and climate in all of history.

It turns out that the Earth's axis of rotation is tilted as it travels around the Sun. The National Aeronautics and Space Administration (NASA) explains that the greater the axis tilts the more the seasons experience dramatic climate changes and the "larger tilt angles favor periods of de-glaciation."[15] When the Smithsonian Institution and New York University researchers looked at the past 800,000 years in the geological record, they discovered that this particular period where hominids evolved was punctuated by the tilt of the earth and sudden extreme changes in the temperature and climate, with repetitive cycles of 100,000 years of glaciation followed by 10,000 years of warming and deglaciation. These dramatic changes in the climate cycle repeated themselves over and over again for 800,000 years and even during the 200,000 to 300,000 years that *Homo sapiens* were coming onto the scene.[16]

How then did our species survive these radical shifts in the temperature and climate on Earth? The researchers' conclusion is that our species is among the most adaptive on Earth. We need to keep in mind that our human family has proven itself to be a wily and adaptive migratory species. Though less physically endowed than many other land and ocean species, we nonetheless survived dramatic changes in the earth's climate and learned to thrive because of our exceptional brain, language skills, the ability to share knowledge and pass it on to offspring and future generations, our opposable thumb and nimble fingers to firmly grasp and manipulate objects of many different shapes to fashion tools, as well as the empathic impulse wired into our neuro-circuitry that encourages collective cooperation.

This study is perhaps the most hopeful note of our times. Our ability to adapt to dramatic changes in the climate of the Earth may still save us, allowing us to flourish in wholly new ways in the uncertain future that awaits us. Each schoolchild should be taught about our species' extraordinary adaptive traits as we face off, cope, and learn how to live on a water planet whose hydrosphere is undergoing a radical transformation.

A partial shift to a migratory and nomadic way of life with shorter sedentary intervals has begun but will likely increase exponentially far into the current century and the future beyond. And, with it, the idea of "attachment to place" will take on a very different meaning. These shorter periods of sedentary life have already taken on a persona – ephemeral cities and enclaves are the new story and the ephemerality of space is fast-becoming the new temporal dynamic of lived experience and a popular buzzword among urban planners, architects, developers, and refugee organizations. Subthemes of the new narrative are proliferating under a constellation of banners, including urban nomadism, liquid architecture, and temporary urbanism, to mention a few.

Ephemeral Cities and Ephemeral Waters

When we think of attachment to place, at least in the Western world, permanence comes to mind. Resilience, especially in urban communities with a long sedentary history, comes with a sense of belonging and at least the subconscious idea of an atemporal space that endures irrespective of the flow of time. The Pantheon, a Roman temple built in 126 CE to honor the deities that ruled over the Roman Empire, speaks to permanence rather than ephemerality. The structure, which is among the oldest intact buildings on Earth, was converted to a Catholic Church in 609 CE and remains the most visited ancient building in Italy.

The Pantheon invokes "eternal being" over the mercurial nature of earthly temporality. Generations of Roman citizens over two millennia of history have visited the ancient temple, comforted by the immortality that awaits them after their short stay on Earth. Similar buildings conjuring up permanence exist across all the great hydraulic civilizations, at least in the Western world. What's interesting is how different the Western approach to religion and philosophy is to the Eastern approach in addressing the question of permanence and ephemerality: or to put it in philosophical terms, "being" vs. "becoming."

The American architect, Jacqueline Armada, contrasts the great divide that has long differentiated Western architecture from the Eastern tradition. She writes that:

> Throughout history, [Western] architects have been concerned with issues of permanence and monumentality, seeking to create and preserve meaning in buildings and the rituals surrounding them . . . [while] architecture of the ancient Western world demonstrates humankind's quest for immortality and godliness through monumental buildings that strive for perfection . . . eastern architecture embraced the impermanence of the natural world.[17]

Clay Lancaster, an American architectural historian, notes that in comparing architectural paradigms in the West and the East, "the first principle that comes to mind is that of solidarity in the West as opposed to fragility in the East. The Western edifice is composed of thick masonry walls as against Eastern construction of slender timbers."[18]

If the Western approach to architecture is one of detachment from nature and autonomy exercised by cloistering, the Japanese tradition emphasizes engagement and embeddedness in nature. Japanese architects have a word for this – *engawa* – which marks the edge between a building and the natural world that surrounds it. Buildings are conceived as semi-permeable membranes that reach out, take in, and emit the natural world. Just as every human being is continually taking in elements and minerals from the hydrosphere, lithosphere, atmosphere, and biosphere, where they reside in our cells, tissues, and organs for a brief period of time, so too, our buildings are semi-permeable membranes. The "edge" is where nature and human habitats intermingle and co-join in a seamless dance – each is embedded in the other in recognition that all that happens on our planet is of relationships manifested in the endless evolution of patterns upon patterns that together make up an indivisible and animated life force.

Kevin Nute, a professor of architectural theory at Cambridge University, makes the point that Japanese architecture, unlike in the West, celebrates the adaptability and even fragility of life via its architectural designs and deployment. The Japanese idea of structure is utterly different from the Western mode. Nute observes that for a Japanese architect, structure:

> acts as a mediator between the human logic of the horizontal floor plane and the natural lie of the land. The two orders coexist in close proximity and in doing so help to define one another. Again, there's a real sense in which the built form seems to belong to the place, the identity of which, far from being destroyed by the building, is actively reinforced by its presence.[19]

The American architect, Frank Lloyd Wright, took a similar approach to attachment to place with his much-praised architectural *tour de force*, Fallingwater . . . a home embedded in a waterfall. Wright's building was designed and deployed to reflect our species' integral harmonization with nature and Wright admitted that his life's work owed much to the ancient tradition of Japanese architecture. The Japanese architect, Tadao Ando, after visiting Fallingwater, opined that "Wright learned the most important aspect of architecture, the treatment of space, from Japanese architecture. When I visited Fallingwater in Pennsylvania, I found that same sensibility of space. But there were the additional sounds of nature that appealed to me."[20]

Nowhere is the contrast between Western and Asian approaches to architecture more vivid than how they view seasonal changes in the weather. In the West, at least in the twentieth century, our habitats have been designed to maintain a conditioned room temperature of approximately 70 degrees Fahrenheit to optimize comfort in an enclosed milieu, independent of and sealed off from the seasonal changes in weather patterns. Japanese, Chinese, Korean, and other Asian structures, by comparison, act more as semi-permeable membranes cued moment-to-moment to the changes in weather and the passing of the seasons. Matsuda Naonori, an Asian urban planner, observes that Japanese architects place a premium on buildings that breathe with thin envelopes that intimately connect the inside with the changes in the weather and the passing of the seasons. He explains that "Japanese people actually preferred the seasonal changes of weather and vegetation in spite of the associated physical discomfort during extreme weather. This situation is one that continues to puzzle Western observers."[21]

It's true that the Asian urban architectural tradition favors built environments that are embedded in nature's ever-evolving processes, patterns, and flows, accompanied by the recognition that our species, like all others, is an intimate part of nature and must continuously seek harmony with the natural world of which it is part. In practice, it's not always cut and dry.

Japan, for example, has swayed back and forth between two approaches to architecture, infrastructure, and urban planning. Both speak to different ways of addressing resilience. The first mirrors the Western approach with the emphasis on "strength," "resistance," and "bounce back" from climate shocks and, in Japan, particularly massive earthquakes and tsunamis. The second approach reflects the Eastern view of resilience as "flexibility," "adaptivity," and "re-generativity" with an emphasis on pragmatic adjustments to the planet's ebbs and flows and continuous surprises. We do so by embedding our species into the fabric of nature's ever-evolving twists and turns, all the while knowing there is no failsafe way to secure life on a hyper-alive and boisterous planet that is forever morphing in unexpected ways – yet, at the same time, realizing that our best bet to survive and flourish is to use our special gifts of language, technological acumen, empathic drive in our neurocircuitry and biophilia consciousness, which favors cooperation with nature and, by so doing, secure our well-being.

The Western architectural tradition, then, emphasizes space detached from temporality, clutching to the idea of an eternal world of pure being unmarked by the ravages of time. The Eastern religions approach the built environment in a more nuanced way, accepting the impermanence of existence and the passage of time, while holding on to a more animist sense of what constitutes immortality. Writing on the fragility of oriental architecture, Lancaster points out that "Buddhist doctrine is permeated with awareness of the impermanence of physical phenomena. Objects disintegrate and individuals die, and although their components continue to exist, the thing itself has ceased to be."[22]

Eastern religions make room for a different kind of immortality – one far closer to how our physicists, biologists, chemists, and ecologists have come to decipher existence. The architect, Günter Nitschke, cites the example of the Ise Shinto shrine in Japan. The building celebrates the death and rebirth of life much the same way our forager/hunter ancestors did with the changing of the seasons and the death and rebirth of each yearly cycle. Every twenty years, a new shrine is built alongside the existing shrine, after which the previous structure is taken down. Twenty-year intervals celebrate the approximate timeline that marks the coming of each new generation. The replacement of the shrine has gone on without lapse for centuries so far into the past that there is no record of when the shrine was first constructed.

How unlike the Pantheon, which stands majestically in space and is seemingly immune to the passage of time. The Pantheon's resilience conjures up a sense of immortality. The Eastern approach to resilience, by contrast, is more accepting of change and bends with the wind. The latter is far closer to the experientiality of nature as fluid processes, patterns, and flows, while the Western tradition, for the most part, experiences nature as passive atemporal objects, structures, and functions.

Ephemeral architecture and its various offshoots are readying our species for a neo-animist future – one far more sophisticated and steeped in what our scientific community is now learning about the ephemerality of existence on Earth. Ephemeral architecture, however, is as mindful of the past as it is of the present and future. The new ephemerality pays homage to memories, celebrating attachment to place as a temporal experience. For example, the new ephemerality often includes the practice of spolia – incorporating fragments of the past into new ephemeral habitats often embedded in the walls, ceilings, and floors as a constant reminder of the temporality of nature that makes up one's cultural heritage. Spolia might be an ancient ornament, or a stone taken from a former shrine or even a fossil once alive in a past geological era. Spolia is a way of experiencing existence as a temporal journey – a continuous ever-evolving and morphing reality where no moment is ever truly lost but lives on and ripples out, affecting every moment of existence yet to come. The point is, there is no such thing as a trivial experience – all is part of the unpredictable choreographing of existence.

Various shades of ephemeral urbanism are mushrooming. "Temporary urbanism" is among the newcomers and has been gaining traction over the past decade or so. It describes initiatives that use unoccupied land or buildings to revitalize urban life for a brief period of time before more orthodox development sets in. Often these vacant spaces and empty buildings remain unoccupied for years at a time and become a kind of "no man's land" and prey to street crime and deterioration. If these neighborhoods remain closed and shuttered for any substantial length of time, they devalue the neighborhood and become an eyesore and a detriment to the entire urban complex.

Temporary urbanists take possession of the buildings and vacant lots without land titles, often transforming the sites and structures to establish vibrant ephemeral neighborhoods bustling with a full repertoire of civic, cultural, and economic activity similar to what one might expect in a conventional urban environment but absent master planning and all of the restraints and codes, regulations, and standards that come with overseeing traditional market-driven development. These bottom-up rather than top-down approaches to urban life are more spontaneous and generally co-governed as sprawling commons, often with informal oversight. The communities host street theater, concerts, art festivals, and makeshift sport sites, alongside mobile retail space, often with food trucks, parks and gardens, cooperative housing, and temporary office space, all catering to a more nomadic and itinerant populace.

The current reawakening of ephemeral architecture and ephemeral habitats borrows extensively from past human migrations and resettlement. The tens of millions of migrants, who flooded to America in the nineteenth century and particularly after the Civil War, were enticed to move out across the great North American frontier and settle along the newly established railroad depots, creating whole new cities overnight and from scratch. The American newspaper editor and founder of the *New York Tribune*, Horace Greeley, led the charge in 1865, calling on the new wave of migrants to "go west young man." The U.S. government sweetened the pot with the Homestead Acts, ceding ten percent of all the public land in the United States to 1.6 million homesteaders, mostly west of the Mississippi River, if they could establish legitimate residency – in other words, sequester the land and show occupancy, which translates into "attachment to place."

Often missing in these mass migrations were the construction materials needed to build homes across the frontier. Sears Roebuck – the first great catalogue company – similar in some respects to a paper version of the internet today – made available prefabricated homes in a giant kit that could be shopped in their catalogues and purchased by mail and transported by train with all the lumber, parts, and instructions needed to erect the building. These do-it-yourself house kits could be assembled in a few days and, while many of them were short-lived, a good number of these houses are still operable but have continually evolved with additions and alterations over the years.

Ephemeral cities are inevitable. They are already here. The United Nations High Commissioner for Refugees says that an average of 21 million people were forcibly displaced annually by climate-related weather events over the past fourteen years and that there could be as many as 1.2 billion climate refugees by 2050. Refugees are already fleeing dry and arid lands and parched earth with temperatures ranging between 100 and 122 degrees Fahrenheit, especially in the Middle East and North Africa but also Central and South America and elsewhere. They are picking up their belongings and trekking hundreds and even thousands of miles in search of survivable environments and along the way they find themselves in temporary shelters – refugee camps. Many of these camps, which can house tens of thousands of people, have become ephemeral cities, housing climate refugees for years at a time.

Granted, it's painful to wrap one's head around the idea that by 2050, 4.7 billion people will live in countries with "high and extreme ecological threats" – that's half the world's total population facing possible migration in the throes of a warming climate.[23] They will have no choice. To say we're not ready for a disruption of this magnitude would be an understatement. But the mass migration has already begun and it's both irreversible and our species is unprepared. To begin with, the whole notion of modern statehood with its underlying practice of sovereign governance and enclosed corridors long fought over in the name of geopolitics is springing leaks on every side and soon will be washed away in large parts of the world, as our collective humanity takes to the road and across the mountains and valleys and open seas in search of an accommodating climate. Already, climate refugee camps stretch across entire geographic regions, particularly in the Mediterranean, but also in parts of Asia.

Although in official circles these ephemeral cities are still looked upon as temporary refuges, in most instances they are far more. We pretend that these huddled masses will at some point in time pick up their meager belongings and continue their trek to some unknown future haven. The truth, however, is that generation upon generation of families live out their entire lives in these domiciles . . . they become temporary permanent homes. Millions of climate migrants find themselves in limbo, stateless and not able to return to their native homeland or be free to migrate somewhere else.

Understandably, no government is willing to call these camps prisons. Yet, they carry many of the hallmarks of incarceration. These waystations are often pejoratively referred to as "permanently temporary." The problem is that the international agencies and humanitarian organizations that administer these refugee camps see their mission as securing the survival and health of their residents, but not empowering the inhabitants to act as citizens and manage their affairs, thus denying them agency over their lives in the camps. The overseers – though well-intentioned – see their role more as custodial. That said, the stateless individuals and families are denied the ability to improve their conditions and develop an attachment to place as home, having to accept the reality of being in perpetual limbo.

Despite the socio-political shackles, residents of refugee camps have found imaginative ways around the protocols and painstakingly created informal "commons governance" while in custody. Although stateless in the traditional sense, and likewise incarcerated in large and often remote fenced-in regions, refugees have learned to become increasingly self-governing in the way they live and, though hardly free, they have succeeded in creating greater agency over their lives and their futures. These are new kinds of long-term ephemeral communities that bear the signature of "do it yourself" commons governance, whether it be building up rather than out in overcrowded camps to accommodate their extended families and social life, or establishing businesses, markets, schools, clinics, and hosting sporting events and festivals.

Researchers conducted a detailed study of life in the Za'atari refugee camp, the largest such camp in Jordan and one of the largest in the world. The Jordanian government opened up the camp to Syrian refugees in 2012, anticipating a population of around 20,000 individuals. By 2015, the refugee population had ballooned to more than 83,000 inhabitants distributed across twelve districts.[24] Each district contained the essential streets, neighborhoods, and shelters. The population was equally distributed between men and women, while young people made up 57% of the population, 19% of whom were under five years of age.[25]

There are 32 schools in the camp, 58 community centers, and 8 health facilities, along with 26,000 pre-fabricated shelters, all with the minimal signature of an urban environment. The camp also includes 3,000 informal shops operated by the refugees and more than 1,200 laborers who work for the camps' NGOs. Water is trucked into the camp and the entire facility is electrified, although the power is spotty and sewers are often in disrepair.[26]

While these facilities make up the essential public infrastructure necessary to maintain a viable urban community, what's missing from the official oversight are incentives and the accompanying codes of conduct that allow the indentured the freedom to co-create the rudiments of public life that we normally associate with a vibrant civil society and attachment to place. Unfortunately, the inhabitants are often discouraged, and outright denied agency to make of the environment a sense of belonging – the most essential feature of attachment to place. Again, that's because the protocol of all refugee camps is to provide "temporary" asylum, when the reality is that the inhabitants and their heirs often remain there across generations.

What's most interesting, however, is the informal route by which the refugees transformed the Za'atari camp into a vibrant community of their own making, despite the incarcerated approach and deployment of formal protocols in the establishment and conduct of the camp. For example, the camp, like others, is laid out as a flat geometric grid. But, over time, the social interactions of the refugees have informally transformed the grid by the way they have come to socialize, shop, visit relatives, play, and the like. Even small add-ons like renaming streets, planting trees, tending small gardens, creating a public square to share gossip, promenade, and listen to music with neighbors, foster an informal social agency and create new ways of remaking space into a lived environment with many of the characteristics of a vibrant civil society and, along with it, a kind of informal commons governance. The value of these vibrant social spaces cannot be over-emphasized. It's these milieus that are so essential to fostering relationships and shared memories. Even the smallest additions, like renaming a street, take on meaning and create a sense of agency that brings people together. For example, in the Za'atari camp, the refugees renamed the most active street where people meet, greet, and gossip the "Champs-Élysées," making it their public square.

Let's understand the nature of the crisis and the timeframe moving forward. According to an internal United Nations Human Rights Council (HRC) report, "the average time refugees spend in camps is seventeen years," neither able to return to their homeland nor be taken in by a host country nor allowed to resettle somewhere else. For all intents and purposes, they are stateless. And, with each passing year brought on by a warming climate adding to the list of homeless detained persons, it's estimated that by 2050, "three billion people will be living without access to adequate shelter" – a 200% increase over three decades.[27] Try to imagine the sheer scale of millions upon millions of human beings – whole communities – fleeing lands devastated by atmospheric rivers, floods, droughts, heatwaves, wildfires, hurricanes, and typhoons, making their way to unknown safe havens across an entire planet. Although the current situation is forcing the dispossessed of the Earth to pick up their belongings and take to the road or the sea, all agree that provisions on a scale previously unimaginable will need to be brought into play lest we face Armageddon. All of this is to say that we will need to rethink how we prepare for mass migration punctuated by longer periods of nomadic treks and shorter periods of sedentary life.

To state the obvious, refugee camps ought to be reconsidered as ephemeral cities. And, while humanitarian oversight is valued, these would-be communities should be viewed less as refugee camps and more as ephemeral cities – waystations established as governing commons, ceding substantial authority to the thousands of individuals and families who might well spend their lives in these quarters and/or be forced to migrate in search of more hospitable habitats as they continually adapt to an ever-rewilding hydrosphere.

These climate-driven refugee camps are telling us that the new nomadism is going to follow a now unpredictable hydrological cycle – it's as simple as that. What we haven't yet come to grips with, however, is that political boundaries, national loyalties, and citizenship papers binding our species to specific geographic domains will become increasingly less relevant as entire swaths of the planet become unlivable, forcing hundreds of millions and even billions of human beings to take to the road. The climate scientists are mapping out the direction of the journey: our species is heading due north from the subtropics and mid-latitudes toward the North Pole. This is not a guess nor a prediction that might unfold far off into the future but is already in the here and now. By the end of the century, the idea of being locked down inside arbitrary political boundaries stretching across biomes and ecosystems that can no longer support our species will seem like a distant memory.

Any talk of ephemeral art, ephemeral cities, and ephemeral bioregions is less than complete if not taking into account the role that ephemeral waters play in determining agency across the planet and particularly their impact on human habitats and urban life.

The term "ephemeral waters" is a bit obscure and little-referenced beyond the hydrological sciences. Still, it is a critical component of the hydrosphere that animates life on our planet. Although, as mentioned earlier, a vast amount of water is locked up below the Earth in its mantle and core, the water we rely on is forever flowing on the surface. Many of the rivers, streams, ponds, and lakes across the planet are intermittent and ephemeral, meaning they ebb and flow and even run dry for a time (there are, of course, permanently flowing rivers, at least until lately, but even that is changing as the major rivers of the world – the Euphrates, Tigris, Nile, Amazon, Mississippi, Yangtze, Rhine, the River Po, the Loire, and the Danube, to name a few, that have flowed for as long as our species has inhabited the Earth, are drying up in real time).

When ephemeral waters momentarily re-emerge, they bring to life long-dormant vegetation in a blistering display of color and glory, only to disappear as fast as the ephemeral rivers dry up. Ephemeral streams, ponds, rivers, and lakes are often found in semi-arid or arid regions. Sudden torrential bursts of rain pour down on parched lands, without warning, creating ephemeral rivers and lakes and submerging everything in their path, only to disappear sometimes in a matter of days or over a period of weeks and months.

Let's look at Death Valley in California, arguably the hottest place on Earth. One would not expect a deluge in the driest region of the planet. Yet, it's happening with increasing frequency as the climate warms. The hydrological cycle is slamming the region with downpours. In 2015, a sudden torrential storm rained down on the parched desert and in a matter of a few short days created an ephemeral lake, winding along ten miles in Death Valley Park. The parched soil in Death Valley is so compacted that it is unable to absorb waters, which can remain on the surface for an extended period of time, causing what's called a super bloom – a sudden eruption of desert flowers. Death Valley became transformed into "Life Valley," with a blooming of flowering fields, including yellow and white primroses, orange poppies, and purple sand verbena. But the blossoming of life, like the sudden rush of the waters, is an ephemeral event that disappears as fast as it arrives.

Ephemeral waters will increasingly be the norm rather than the exception in the decades to come, reminding us of the ephemerality of life itself.[28] Intermittent ephemeral streams, ponds, rivers, and lakes play a prominent role in ecosystem dynamics, and their reshuffling is both a troubling reminder but also a hopeful sign of the animating power of our planetary hydrosphere in nurturing life anew. The unexpected coming to life of a magnificent garden atop a barren desert in Death Valley speaks to the majesty of the hydrosphere as the prime mover of life on our home in the universe.

It's hard to realize that the world we live in is being turned upside down and inside out and that many of the markers by which we have come to judge our attachments are but fleeting affairs – what then do we hold on to? For sure, the great reset from sedentary to nomadic life is scary and destabilizing, but coming to the realization that we have misjudged the nature of our existence can also be enlightening and reaffirming; especially knowing that every moment of our individual journeys ripples out and affects all that comes after us. It's the nature of what existence is all about. The point of all of this is that the hydrosphere is now the decider, as it has been throughout the history of our planet, whether acknowledged or dismissed. All of life on Earth and every moment in-between is indeed ephemeral. Existence, by its very nature, is fleeting and conditional, and constrained and empowered, with the Earth's hydrosphere exercising agency at every twist and turn.

Ephemeral cities share common ground with ephemeral waters. By the end of the current century, these ephemeral cities will dot every continent. Ephemeral urbanism is already taking wing and changing the way we think about time and space on an unpredictable planet and where and when we stay and move. Interestingly, the playbook for ephemeral cities comes from unusual quarters – traditional religious gatherings and music and arts festivals. These communal pilgrimages draw hundreds of thousands and even millions of people who constitute a fully functioning city for an ephemeral period of several days to several weeks.

The best known and largest of the ephemeral cities is Kumbh Mela, which occurs every twelve years. The ephemeral city lies on the outskirts of the Indian city of Allahabad, at the confluence of two sacred rivers, the Ganges and Yamuna. Millions of people make a pilgrimage to the religious site to bathe in the holy waters. The ephemeral city, which lies on the floodplain of the two rivers, exists for 55 days and hosts more than seven million residents at any given time and, over the period of the festival, hosts an additional ten to twenty million ephemeral residents. The ephemeral city is erected with clocklike precision over a period of several weeks, but is the result of years of planning and is equipped with roads, electricity, running water, bridges, and tents of various sizes. The facility is overseen by a sophisticated governing body, along with a rich menu of cultural activities designed to create a Commons – all to the end of bringing together millions of pilgrims in consort with each other and the sacred waters of the Ganges and Yamuna rivers. In the weeks after the festival, the infrastructure is packed up, removed, and sent out across India to be refurbished and stored in warehouses during a twelve-year hiatus, in anticipation of the next ephemeral urban gathering.

It is said that the Kumbh Mela pilgrimage and the scores of other such ephemeral gatherings, be they religious or more secular, like the famed Burning Man gathering that takes place over nine days of Labor Day weekend each year in the desert of Northwestern Nevada, come with a second underlying theme. These ephemeral mass gatherings allow precious moments for millions of people to come together on a great Commons stripped of all the superficialities and distinctions that separate one from the other and dwell in communion in deep appreciation of the transitory nature of existence that binds us together to the life force on Earth.

A New Business Model: Additive Manufacturing and Provider–User Networks

While ephemeral cities like the one at Kumbh Mela rely on conventional building materials that can be disassembled, stored, and reused, a digitally savvy generation of architects and urban planners are using the new algorithmic technology of additive manufacturing – 3D printing – to build homes, offices, and other commercial structures in a fraction of the time of conventional construction and then partially disassemble and reuse them elsewhere with short notice as our species becomes increasingly nomadic and less sedentary. The additive 3D printing process for manufacturing differs in both degree and kind from what architects would refer to as subtractive manufacturing, which was the dominant mode of industrial life in the nineteenth and twentieth centuries. "Subtractive manufacturing removes material to create parts" while "additive manufacturing builds objects by adding material layer by layer."[29] The 3D printing of a house begins with a computer program that develops a digital model of the house. "The 3D printer [which is a robot] then prints layers of concrete and these layers are arranged in rows per the design of the model. The process continues until all the rooms, walls, and other concrete-made sections of the house are built."[30]

In 2021, Habitat for Humanity, which builds homes for low-and-moderate-income families, teamed up with the 3D printing company Alquist and built out a 3D printed house – the first owner-occupied 3D printed home in the state of Virginia. All of the exterior walls of the three-bedroom, two-bath owned home were printed out layer by layer in 28 hours – reducing the conventional construction time by more than a month. Because additive manufacturing does not shave off or cut down material in the construction, as would be the case using lumber, bricks, tiles, and other materials, the buildout saved an estimated 15% per square foot in the overall construction cost.[31] The Habitat for Humanity home even comes with its own personal 3D printer, which allows the family to print cabinet knobs, electrical outlets, and numerous other components and features that need to be replaced.[32] The home is also EarthCraft certified, meaning it generates less waste in its construction, uses less energy to warm and cool the house, is more efficient in conserving water, saving on the utility costs, and is more resilient to climate-related disasters.[33]

The developer, Alquist 3D, has announced that it will build 200 more affordable 3D printed homes in Virginia, ranging in price from $175,000 to $350,000 dollars.[34] Alquist CEO Zachary Mannheimer notes that fast-assembled and cheap 3D printed homes that are energy efficient will increasingly become the standard in home construction, especially "with migration patterns shifting due to pandemic, climate, and economic concerns."[35]

3D printed buildings are just the beginning of a radical departure in the deployment of the built environment. GE, for example, has built the biggest additive constructed facility to 3D print the base for wind turbines.[36] In the Netherlands, 3D printing has been used to construct a bridge and elsewhere 3D printing has been used to print out bus stop shelters, public restrooms, rocket engine components, and fuel tanks, to name but a few of the applications.[37] The Spanish company, Iberdrola, in a collaboration with Hyperion Robotics, is using 3D printing to construct the foundations for electricity grids that will transmit renewable green electricity over power lines. The 3D printed foundation blocks use 75% less concrete than conventional subtractive manufacturing.[38]

Other global corporate giants including the Mexican company CEMEX, the fifth largest global materials company, the Swiss chemical giant, Sika, and Saint-Gobain, one of the world's largest construction materials multinationals, are diving into the 3D printing building market. 3D printed additive construction got a boost from the Middle East, where the United Arab Emirates announced its 2030 Vision and Saudi Arabia unveiled its NEOM project. Dubai seeks to make 25% of its buildings 3D printed by 2030, while Saudi Arabia has announced that it will inject $500 billion for planning and construction of 3D printed buildings from the Public Investment Fund of Saudi Arabia and international investors.[39]

Although 3D printing for structures of all kinds uses far less cement, the cement industry still contributes eight percent of the world's CO_2 emissions.[40] Early adopters in the 3D printing field, like Stephan Mansour, suggest that alternative base stocks that are greener are coming online and include "metakaolin, adobe, limestone, recycled construction waste, mine tailings, shales, and more. Similarly, new reinforcement materials will be explored, including hemp and hemp rebar, graphene, embedded fiber, and glass aggregates" and will likely become more viable in the very near future, eliminating cement as a feed stock. Even soil is being used in experiments as potential feed stock.[41]

In 2021, the Italian architect Mario Cucinella, poured out the first clay house sourced entirely from locally available clay soils using 3D printing. The eco-sustainable structure was poured by the printer in 200 hours, with little waste or scraps generated in the construction. The architect uses timeless design principles found in climate resilient ancient structures combined with high end digital printing in the construction. Aside from using locally sourced clay soil, the beehive type envelope of the building with two circular skylights is nearly a zero-emission structure greatly reducing the cost of construction while producing a low carbon house.

The furnishings in the building, which include a living zone, kitchen, and a night zone, are "partly designed to be recycled and reused" and every feature of the design is organically fitted to the local climatic conditions "to balance thermal mass, insulation, and ventilation," according to Cucinella. As for the importance of this breakthrough climate resilient architecture, Cucinella says that what motivated the firm is "the need for sustainable homes . . . and the great global issue of the housing emergency that will have to be faced – particularly in the context of the urgent crises generated, for example, by large migrations or natural disasters."[42]

The software for 3D printed buildings can be licensed via a "provider–user network" instead of a traditional "seller–buyer market" by uploading and instantaneously sending the software instructions at near zero marginal cost to any part of the world, allowing developers onsite to print out buildings on a just-in-time and just-in-need basis and pay a fee to the provider for each building downloaded. This is an example of how the emerging ephemeral society makes possible a new economic exchange paradigm that takes the economy from globalization to glocalization, favoring high-tech small and medium-sized enterprises (SMEs) engaged in a rich latticework of economic exchanges across industries, while avoiding the steep cost of ocean or air transportation and logistics.

While the 3D printing industry is finding ever greener sources of feedstock for constructing resilient buildings and all of the other structures necessary for assembling pop-up ephemeral cities in a climate ravaged world, attention has also turned to the second aspect of making climate resilient pop-up cities viable ephemeral habitats. Increasing attention will need to be focused on the ease in disassembling, refurbishing, recycling, transporting, and reusing many of the components of the buildings in new sites and locations where 3D printed buildings – commercial and residential – can be quickly poured, creating new cities to accommodate a more nomadic population in an ever-evolving migratory world.

IKEA, the world's largest retailer of home, office, and industrial furnishings, is leading the way in introducing ephemerality into its more than 9,500 product lines. The goal is to make products that can easily be disassembled, refurbished, repaired, recycled, and resold to other users, creating a wholly circular economy with minimum waste. In this new economic approach, IKEA is even beginning to make a partial transition from traditional seller–buyer markets to provider–user networks. The global company is shifting the very way the economy works. In provider–user networks, markets partially give way to shared commons in the exchange of goods and services.[43] The IKEA approach allows for about anything inside a home, office, or factory to be ephemeral and mobile so they can easily be disassembled, reworked, and reused in new locations by an increasingly nomadic population, which is continuously migrating to new regions to adjust to the changes in the planet's hydrological cycle caused by a rapidly warming climate.

The ephemerality of existence is not a theory but the way the universe has operated since the Big Bang 13.8 billion years ago. The first and second laws of thermodynamics rule our existence everywhere in the cosmos, including our little oasis on Earth. Plato and all of the footnotes to his philosophical musings, stretching across history to Immanuel Kant and the philosophers and scientists of the European Enlightenment, would have us believe that the temporality of existence is a mirage or, at best, a poor imitation of a higher state of pure being in the otherworld, where rationality rides above corporality and is an immortal force ruled by mathematics and pure forms that are without a physical face. This is how many of our Western philosophers have come to define the essence of existence.

Our common sense, however, over the eons of time in which we've inhabited the Earth tells us that these philosophical oddities are strange and even eerie. In a sense, there's a strong current, especially in Western philosophy and religious traditions, that continues to curl up around a fictional reality – a utopic otherworldly existence that disparages the physicality and ephemerality of lived experience – frightened of the prospect of the beginnings, passing, endings, and rebirths of the life force on Earth. But try as they will with ever-more arcane explanations favoring being over becoming, our everyday experiences tell us that all of life is ephemeral.

Bringing this down to Earth, a new generation of physicists, chemists, biologists, ecologists, artists, sociologists, psychologists, and anthropologists are rediscovering the ephemeral nature of existence and are beginning to revisit how our ancient ancestors perceived the world but through the lens of a new more sophisticated animism that goes apace with the collapse of hydraulic civilization and the rise of ephemeral society. On Earth, as it is in the universe, all is ephemeral, and each experience leaves an indelible imprint and lives on with agency, affecting every phenomenon that follows – what might be considered an ephemeral immortality.

Now, the sixth extinction of life on Earth is speeding the planet's great spheres – the hydrosphere, lithosphere, atmosphere, and biosphere – into ephemeral overdrive to find a new forward that will keep the torch lit for life to flourish in different ways. Our species, like our fellow creatures, is waking up to a new ephemeral journey in lock-step with the four spheres that animate earthly existence on the planet. We are returning to our biological roots – our nomadic inheritance – with a much more enlightened understanding of how our very existence is gestated and maintained by the waters of the Earth as it is with all other life. That reawakening in the form of a new mindful, astute, and scientifically driven neo-animism is already expressing itself in the arts and in ephemeral architecture and urban planning.

This rebirth comes with a new sense of our species' intertwined relationship with the Earth's hydrosphere. Ephemeral cities will be at the heart of the new nomadism. This great ephemeral reset is already sweeping our species into a new framing of both governance and economic life. All in all, we are coming to have a more enlightened understanding of how to steward rather than simply manage and extract the Earth by treating the planet's other powerful agencies as a "global commons." This reset will require exercising the empathic and biophilic roots that are ingrained in our neural circuitry in the exercise of governance of local ecosystems. Mapping out our new relationship to the planet's biomes and ecosystems begins with rethinking every aspect of how we dwell on Earth alongside the other agencies on the planet.

CHAPTER 10

Bringing High-Tech Agriculture Indoors

Nowhere will the effects of freeing the waters be more consequential than on how our species feeds itself. The United Nations Foundation cautions that global food yields could decline by as much as 30% by 2050, setting off mass hunger, starvation, and death, accompanied by an historic migration of human population.[1]

Some of the countries that produce much of the world's foods – including the United States, China, India, Australia, and Spain – have either reached or are close to reaching their limit in food production because of the dwindling amount of available water. A case in point: the Aral Sea in Central Asia, once the fourth largest freshwater lake in the world, lost a volume of water equivalent to all of Lake Michigan in three decades because much of its waters were diverted to irrigation and power generation. As the sea diminished, it left behind polluted land and the region's stranded population with acute food shortages, an increase in infant mortality, and a decrease in life expectancy.[2]

Unfortunately, America turned a blind eye to the tragedy that unfolded in the Aral Sea, only to follow suit with an even bigger calamity. The great Colorado River, which has watered much of the Western United States for nearly a century, is already three-fourths empty, and parts of Lake Mead are drying up, exposing the eerie remains of human cadavers and crashed automobiles from decades past. As mentioned earlier, the Colorado River runs into Lake Mead and from there over the Hoover Dam.[3] The dam is now heading to obsolescence. Both the

Colorado River and Lake Mead are spiraling into a death throe, threatening the survivability of the western United States with a population of over forty million Americans.

According to Abrahm Lustgarten, a senior environmental reporter at *ProPublica*,

> the suggestion by the federal government is that states agree to cut between two and four million-acre feet [of water] in 2023. And acre feet is . . . the volume of water used to measure the Colorado. But, just to give a sense of scale, the river is running at about nine to ten million-acre feet now and even in its best days ran at about twelve to thirteen million-acre feet. So, we're really talking about . . . cutting 40 to 50% of current use on the Colorado River.[4]

As the federal government begins the process of imposing mandates,[5] this would mean either dramatically reducing agricultural production across California and the American west, and/or severely reducing water for industry and for human use, forcing the largest mass migration in U.S. history.

Notwithstanding the dire signs, population in the West continues to increase as retirees and even a younger generation from around the country flood into the region, taking up residence in newly developed condominium communities across the desert, with names like "Desert Shores" and "The Lakes." The *New York Times* columnist Timothy Egan said that if we "think of Lake Mead as the world's largest heart monitor . . . it's showing extreme distress."[6] With climate change triggering record-breaking temperatures across the arid West, the inevitable comeuppance is upon us, but has still not been fully absorbed, as new golf courses continue to be laid down.

Some residents and businesses are already pressing for establishing a continental pipeline that can suck fresh water from the Great Lakes – the largest body of fresh water on Earth – and transport it via pipelines to the western U.S. to keep the states viable. If this state of affairs isn't enough to come to a reckoning over climate change, the warming of the planet, and a revolt of the waters, it's difficult to know what it will take to jar public consciousness in time to start the great migration out of the west and upwards and eastward on the North American continent following the lead of the hydrological cycle. The lesson here is that despite all the prowess that allowed our collective humanity to believe it could corral the planet's hydrosphere, it has always been a fool's errand.

To get a sense of the way water is indiscriminately used, consider this. There are ten million acres of lawn in the U.S., which require 270 billion gallons of water every seven days. That water could provide all 7.8 billion individuals on Earth with a shower four days each week.[7] With the loss of the Earth's topsoil, less available arable land, water shortages, and a hydrological cycle that's radically changing the distribution of water on Earth, our species will need to rethink how we secure and use the waters.

Vertical Farming

Adapting to the freeing of the waters so they can run their course and rewild the landmasses of the planet to spark the rejuvenation of the soils, reforestation, and a flourishing of life will likely mean growing an increasing amount of our food indoors, using a fraction of the water required in conventional outdoor agriculture. Till late, the idea of indoor agriculture would have been viewed as far-fetched – not anymore.

Stunning technological breakthroughs in growing food indoors in the past decade are taking our species into a new era that redefines the very way we think of securing food. It's called vertical farming or indoor agriculture. With indoor farming, plants are stacked vertically in layers in giant warehouse-like facilities equipped with large shelving units. The first thing to know about vertical farming is that it doesn't use soil but instead chemicals mixed in water to stimulate growth. The second thing is that the plants are never exposed to sunlight. LED lighting substitutes for the sun in assisting the process of photosynthesis and robotics, digitized surveillance, and algorithmic governance determine the plant's nutritional requirements to assure proper growth and maturation. The system is designed to control temperature, humidity, CO_2, and light to create optimal growth. Because the crops are grown indoors, they do not require the use of pesticides. And then there is the big payoff. The plants use up to two percent of the water that a conventional outdoor farm uses, and that nutrient rich water is continually recycled in an internal irrigation system, meaning that less water is used throughout the growth process.[8]

Since vertical farms grow plants upwards in facilities instead of in horizontal rows in fields, they take up far less space, increasing their productivity per acre. Additionally, because they are grown in warehouse-like structures, vertical farms can be located near population centers, cutting transportation costs and ensuring that the produce can be delivered to the supermarkets and restaurants in a matter of hours rather than days, weeks, and months, saving time and fuel. Vertical farms can also produce up to fifteen harvests per year, compared to conventional outdoor farms that harvest once or twice a season.[9] An additional value to vertical farming is that it doesn't require onsite management and inspection. Remote surveillance monitored by algorithmic oversight continually oversees the plants and adjusts to changes in real time, affecting the plants' growth. Such facilities are likely to be the future of indoor farming.

Indoor vertical farming is also rapidly expanding to include plant-based meats. Plant-based meat was first introduced in the market as far back at the nineteenth century. Today's leading plant-based meats – beef, chicken, pork, etc. – were introduced in the 1970s and have grown in popularity ever since. Veggie burgers are now commonplace and marketed by the world's leading agricultural giants including Tyson and Nestlé.

A younger generation, concerned about the enormous carbon footprint and water footprint that comes with eating beef, pork, chicken, and other animal products, not to mention the horrific toll in animal cruelty and public health in consuming a meat-based diet, has increasingly veered to a more vegan dietary regime, particularly in wealthier countries and global cities. Still, this new approach to cuisine is more nuanced than at first glance and requires a sophisticated knowledge of the ingredients that go into plant-based burgers, hot dogs, chicken fillets, and the rest. While conventional grains – wheat, rice, etc. – are a common food stock for plant-based meat alternatives, they come with a more elevated carbon footprint. Legumes – chickpeas, black eyed peas, pinto beans, and edamame – are equally high in protein, but with smaller carbon and virtual water footprints and are a healthier and more climate-friendly alternative. But even here, consumers need to be mindful of all the other additional ingredients that are often included in the plant-based meat alternatives like corn sugars and equally troubling canola oil that contains trans-fats. As indoor vertical farming commands increasing market share, plant-based meats are expected to win over a younger generation around the world.

While indoor high-tech laboratory generated sourcing of conventional foods is a growing market that's going to scale exponentially and spread globally over the coming decades, a new even more dramatic change in provisioning food is afoot that's likely to alter food habits for millennia to come, as the planet's hydrosphere renegotiates the rewilding of the Earth. In July 2021, the World Economic Forum, the premier global platform, think tank, and incubator for new ideas to spur the economy, chose to run an article penned by Antoine Hubert, President and Chief Executive Officer of Ÿnsect. His company promotes a transformational shift in the diet to meet the needs of a human race that is careening to 9.7 billion people by 2050, despite "only four percent of arable land remaining available on the surface of the planet."[10]

The new menu he proposes is insects, not as an occasional swallow but as a central part of the human diet ... not all that new an idea. Entomophagy is the term used for humans eating insects. Two billion people today, mostly in Asia, Latin America, and Africa, already include insects in their diet. The practice dates far back in time to the early beginnings of our species' existence on Earth.

Up until now, however, the contributions that insects have made to the human diet have been at most complementary, rather than a mainstay, because the other sources of food were more abundant, available and easier to harvest, store, and conserve in sufficient quantity to meet dietary needs. Now that has all changed for two reasons. We are running out of arable land to feed our population and we are in the midst of an extinction event that is already eliminating much of the animal life – both terrestrial and aquatic – that has been a centerpiece of our diet.

Insect farming has labored on the margins of food production, often unobserved, for quite some time. Yet, more than a trillion insects are raised on farms each year for human consumption and animal feed. The problem has been that the process is manual and slow and not able to scale at costs that would make it a central part of the human diet. That is now rapidly changing. The introduction of AI and the Internet of Things is making it possible to "industrialize the breeding of insects in a contained environment," just as the Ag industry has begun to introduce precision IoT-directed agriculture and, more recently, the cultivating of vertical indoor high-tech agricultural crops.[11]

Antoine Hubert's company, Ÿnsect, has already more than 300 patents covering a fully automated AI-driven process that is able to produce 100,000 tons of insect products per year. He explains how the process works:

> More than one billion data points (vision, weight, temperature, development, speed, weather, composition) are captured daily and fed into a proprietary predictive model to optimize insect breeding and rearing conditions. Thanks to AI, with just one image taken at the right time, up to 80% of the quality control information for the daily care of the breeding of mealworm insects can be derived. [For example,] the handling of *Tenebrio molitor* insects is entirely done by programmed robots. Machines do the heavy lifting: they fetch the various bins from the vertical farm and bring them to one area to be fed or to another to collect eggs and larvae which, when ripe, are taken to the processing stage.

The North American Coalition for Insect Agriculture is quick to point out the advantages of including insects as a central part of the human diet. The association notes that more than 2,000 species of insects are or have been consumed as food sources – offering up a variety of new food crops compared to the very limited dietary options currently available on the global food menu. These insect food sources include crickets, beetles, moths, bees, ants, wasps, grasshoppers, and dragonflies. A review of the nutritional value of edible insects found that grasshoppers, crickets, meal, and buffalo worms have more iron, copper, magnesium, manganese, and zinc than sirloin beef and are naturally gluten free and compatible with the Paleo diet. As to protein, it turns out that crickets alone "pack more protein than cows, chickens, and pigs." Bringing this home, studies show that "when compared to chicken, one gram of edible protein requires two or three times as much land and 50% more water compared to mealworms" and "a gram of edible protein from beef requires 8–14 times as much land and approximately five times as much water compared to mealworms."[12]

Insects also emit considerably fewer greenhouse gases than most livestock. To cut to the bone, studies show that "crickets are twice as efficient at converting feed to meat as chicken, at least four times as efficient as pigs and twelve times as efficient as cattle when the protein of each animal that is edible and digestible is considered." And, if these stats aren't enough to turn heads, bugs are cold-blooded, meaning that they do not waste energy converting feed into body heat.[13]

While bugs have long been a staple in dietary regimes across Central and South America, Australia, Africa, and Asia, introducing these foods to populations in Europe, the U.S., Canada, and other nations has been spotty at best. But now some of the world's leading celebrity chefs are seeding the ground by introducing entomophagy onto their menus. René Redzepi, a world-renowned Danish chef, has taken a lead in introducing dishes such as crème fraiche dotted with ants at his London pop-up restaurant, while in Japan he has debuted on the menu a French sushi dish crawling with tiny black insects, and in Mexico he has featured a tostada with grilled espolon beans and creamy escomoles (ant eggs).[14] Though new to middle-class Westerners, entomophagy has long been a staple in Asian cuisine. Fried insects are a common snack in Cambodia, as are spring rolls stuffed with ants and green papaya salad with grilled marinated scorpions and makrut lime.[15]

Indoor virtual farming of insects at scale is likely going to boom in coming decades and even partially replace beef, chicken, pork, and other meat-based foods, as successive generations become weaned on the new cuisine at an early age, modifying our species' dietary regime to accommodate both a global warming climate and a more ephemeral nomadic lifestyle – after all, insects are ubiquitous and found everywhere, making this food source cultivated in pop-up vertical indoor laboratories an appropriate fit for a migrating human population. If this dramatic change in diet seems a bit too squeamish to entertain, we ought not to forget that sushi – in the form of raw eel, squid, sea urchins, and scallops – was first introduced into the U.S. in a single sushi bar in Los Angeles in 1966.[16] Today, the sushi market in the United States is valued at $26.6 billion and is particularly popular among millions of young people.

High-tech indoor vertical farming comes with two other advantages. Were a severe climate event to take down power lines and disrupt far-flung logistics and supply chains across regions, landmasses, and even oceans, vertical farms nestled near population centers and equipped with onsite or nearby solar and wind-generated electricity – both utility scale and microgrids – would continue to keep the indoor plant growth on schedule. Were worsening climate disasters to force mass migrations of population, pop-up indoor vertical farm facilities could be disassembled at a moment's notice and reassembled along migratory corridors, providing emergency foods for extended periods of time.

No one in the vertical farming milieu believes that indoor agriculture is going to be able to substitute, *en masse*, for outdoor farming, but there is little doubt that it will play an increasingly important adjunct role providing backup resilience on an ever-more climate-related disaster-prone planet.

CHAPTER 11

The Eclipse of Sovereign Nation States and the Gestation of Bioregional Governance

The warming of our global climate is going to force a rethinking of political boundaries, national sovereignty, and citizenship. We should have seen it coming, at least since the late 1980s when a warming climate began to make its mark, with a steep increase in floods, droughts, heatwaves, wildfires, hurricanes, and typhoons, destabilizing entire ecosystems, wreaking havoc on infrastructure and taking the lives not only of our species, but our fellow creatures as well.

Mass Migration and the Issuing of Climate Passports

The wake-up call is now, and the way we know it is the sudden interest in rethinking passports everywhere – a document most of us pay scant attention to, except in planning vacations and business trips abroad. Passports have suddenly become a vital lifeline – both negatively and positively – depending on what part of the globe one is attached to. Passports were once considered a pleasant luxury of sorts allowing us to visit family and friends abroad and immerse ourselves in different cultures and experiences, including enjoying the rich diversity of nature's gifts. Today, passports are increasingly becoming a signal of either safe havens or menacing new milieus.

The mass migration toward the northern climes is going to topple some national governments across the subtropics and mid-latitudes in coming decades.[1] There will be no holding back either the mayhem or the morphing nomadism. It's a done deal, even now. This doesn't mean that governance is going to disappear overnight in a free-for-all as waves of humanity scatter to the winds and head mostly north. But, already, the tell-tale signs of a great political upheaval are everywhere to be seen. We need look no further than the dramatic changes taking place in the issuing of passports – once the ideological stamp of nation-state sovereignty and the principal way that much of the human race came to define their very existence.

A sort of foretaste of what's occurring now unfolded a century ago immediately after World War I and the armistice, which foretells what's likely to occur in the coming decades as national political boundaries are toppled in the wake of global warming, forcing climate refugees to travel north. The peace armistice signed in 1918 ending World War I left chaos across the landscape. The newly minted Soviet Union revoked the citizenship of 800,000 Russians who fled their country during the war and were living abroad. The collapse of the Romanov dynasty and the seizure of power by the Russian Communist Party triggered a domino effect, with similar collapses in Germany and the Austro-Hungarian and Ottoman empires, leaving a political vacuum. Suddenly millions of people were stripped of their political identity. The vacuum was hurriedly filled with the creation of new nation states and accompanying boundaries. This sparked further crises as minority ethnic groups found themselves attached to governing entities that were less than welcoming. Isolated and stateless, millions of families were cornered and stripped of their previous national identities and unable either to stay or go because they no longer had a formal political identity and had found themselves without a country. In legal terms, they became the invisible other.

It was at this time that a famous Norwegian polar explorer Fridtjof Nansen became the new head of the office of the High Commissioner for Refugees of the then League of Nations, tasked with assisting stateless refugees across Europe. Nansen convened a conference at the League of Nations in Geneva and obtained the consent of the member states of the league to allow him to issue what was called a "Nansen certificate," first issued to Russian refugees and sometime thereafter to Armenian refugees, who made up the bulk of the stateless individuals across Europe. The so-called Nansen passport was valid for twelve months and could be renewed, allowing recipients to travel across nation-state boundaries and find a welcoming home in one or the other of the League of Nations member states.

More than 450,000 refugees used Nansen passports recognized by 52 countries until 1942. The main features of the passport were embedded in the 1951 Refugee Convention, allowing the bearer the legal right to enter any nation that accepted their status, thus incorporating the concept into international refugee law. While the Nansen passport was of great assistance primarily to Russian and Armenian refugees, others in the Balkans and elsewhere were never given this legal entitlement. Still, the Nansen passport drew a line in the sand in international law, legitimizing the right of stateless people to seek a new homeland and be accepted.[2] The Nansen passport opened the door to a rethinking of passports and the whole idea of belonging to only a single country.

The road ahead is challenging as we begin to leave behind a world where everyone is expected to belong over their lifetime to a single fixed geographical space under the aegis of a sovereign nation, albeit with permission to travel abroad for business or pleasure, or simply migrate to another country and give up their original citizenship. On a global warming planet where entire regions of the world are becoming uninhabitable, sedentary life will be less tenable while nomadic life will increasingly become the norm.

Gaia Vince, the British environmental journalist, and author of *Nomad Century*, casts our future migratory journey in both stark relief and a hopeful light. She writes:

> At least as challenging, though, will be the task of overcoming the idea that we belong to a particular land and that it belongs to us. We will need to assimilate into globally diverse societies living in new polar cities. We will need to be ready to move again when necessary. With every degree of temperature increase, roughly one billion people will be pushed outside the zone in which humans have lived for thousands of years. We are running out of time to manage the coming upheaval before it becomes overwhelming and deadly.[3]

Although at first glance one might suspect that passports have remained relatively rigid, locked into an outdated geopolitical frame designed to keep the citizenry locked up. That's not the case. At least since the end of World War II and the coming of globalization of trade, accompanied by a dense interconnected latticework of economic activity populated by transnational corporations and a highly mobile global workforce, passports became far more agile and less restrictive, allowing businesses and labor forces to operate on a global playing field. The globalization of the economy and commerce has been accompanied by new global governing institutions including the UN, the World Bank, IMF, and OECD, which have transformed economic life, making it a planetary phenomenon. Then, too, the globalization of flight opened up the travel and tourism industry and turned the planet into a giant playground, affordable for hundreds and millions of human beings.

Passports changed with the economic winds. Each year the world's passports are ranked from most agile to least agile, based on the number of destinations that don't require a visa from the destination country – which can take weeks or even months to obtain. Japan tops the list. Its passport holders are able to visit 193 nations without the approval of the destination country. Afghanistan currently ranks at the very bottom, with only 27 destination countries not requiring a visa.[4]

Even more revealing is the ever-increasing number of countries that permit dual citizenship, allowing individuals to be citizens of two or even more countries at the same time. Dual citizenship means that an individual has all the rights of citizenship, including the right to vote, seek employment, buy a home, receive all the standard public services including healthcare and public education. Of course, dual citizenship also comes with obligations, including payment of taxes and serving in the military if there is a conscription.

Dual citizenship also eases the coming together of extended diasporas, as ethnic groups increasingly remain connected via the internet, Zoom, and cheap international flights. Living part of one's life, both virtually and physically, in two or more places is a sign of the new semi-nomadism sparked by globalization, commerce and trade, and cheap travel.

The easing of travel visas and the granting of dual citizenship provides a baseline for extending citizenship. With climate change hastening mass migration on a scale without precedent, conversation is quickly moving to the idea of issuing global climate passports to allow climate migrants to cross national boundaries along designated corridors, taking up sedentary residence for a short period in migrant waystations for respite and then to travel sometimes thousands of miles to more climate-friendly regions.[5] On an even larger scale, the European Union comprises 27 different nations and 447 million citizens, each of whom think of their homes as both the EU and their respective state and region. All EU citizens carry an EU passport.

The United Nations has made the issue of addressing the plight of refugees a central theme since its inception in 1945, issuing voluminous studies and reports and enacting legally binding conventions dealing with safeguarding the rights of refugees. Still, the international body, until late, has tiptoed around the question of how to broach the subject of climate refugees – the hundreds of millions and soon billions of human beings that threaten to crash through the political borders of their sovereign nation and take to the road in a desperate flight to find a hospitable climate and friendly governing jurisdiction willing to take them in.

In 2021, the German Advisory Council on Global Change, a scientific advisory body to the German federal government took the plunge – be it a shallow dive – issuing a report that addresses the growing refugee crisis in which they included the question of legitimizing the plight of climate refugees, referring to the earlier Nansen passport which assured safe passage to stateless refugees in the wake of World War I to welcoming by host countries.

The commission stepped forward with a bold proposal, calling on the nations of the world to recognize the rights of climate refugees by issuing climate passports. The commission recommendation is a potential game changer that, if acted upon, would have far-reaching geopolitical consequences, the least being weakening the hold that nation states enjoy with closed borders. The recommendation reads as follows:

> The WBGU therefore proposes a climate passport for migrants as a key instrument of a humane climate policy. Based on the Nansen passport, this document would offer people existentially threatened by global warming the option of having access to – and rights equivalent to citizens' rights in – largely safe countries. In the first phase, the climate passport would open up early, voluntary and humane migration pathways to the populations of small island states whose territory is likely to become uninhabitable as a result of climate change. In the medium term, the passport should also be available to people under massive threat in other countries, including displaced peoples. States with considerable historical and present-day greenhouse-gas emissions, which therefore bear a considerable

> amount of responsibility for climate change, should offer their services as host countries.[6]

Incredibly, the existing UN covenants and protocols do not recognize climate refugees, even though millions of climate migrants are already fleeing their homes in search of other climate-friendly regions to live in. The UN covenants only provide protection against human abuses – for example, government persecution of religious sects, ethnic minorities, and political parties, but not refugees fleeing a global warming climate.

The Geneva Convention, also known as the 1951 Refugee Convention, is a UN multilateral treaty that defines a refugee and establishes the rights of individuals for the purpose of granting asylum to nations other than their homeland. Here's how Article 1 of the convention defines a refugee:

> As a result of events [. . .] and owing to well-founded fears of being persecuted for reasons of race, religion, nationality, membership of a particular social group or political opinion, is outside the country of his nationality and is unable or, owing to such fear, is willing to avail himself of the protection of that country; or who, not having a nationality and being outside the country of his former habitual residence as a result of such events, is unable or, owing to such fear, is unwilling to return to it.[7]

Unfortunately, the German Advisory Council on Global Change recommendations suggesting the issuing of climate passports has to date fallen on deaf ears and been pushed aside by the United Nations. The fear is that it would lead to a mass-exodus from home countries, with hundreds of millions of climate refugees wandering across continents and oceans in search of a host country in a more amenable climate milieu. The UN has sidestepped the growing crisis, asserting that countries ravaged by climate change be provided with assistance in the form of loans to bolster internal resilience to climate impacts – all necessary, but belies the reality that whole regions of the world are already becoming unlivable, and millions of climate refugees have taken to the road in search of more benign climates.

Although the German Advisory Council on Global Change recommendations for a climate passport was essentially dead-on-arrival at the UN, the global body did put on the table the Global Compact for Safe, Orderly, and Regular Migration for consideration. While the compact listed climate change as a cause for flight and called for cooperation by the member states, even this light touch fell by the wayside as sovereign nations were concerned that, should climate passports be part of the package, it would lead to a mass migration of climate refugees from all across the world. Even the German government gave a thumbs down to the recommendation from its own Advisory Council on Global Change, assuring the UN member states that the Global Compact for Safe, Orderly, and Regular Migration should not become a legally binding agreement, but only an aspirational accord, or what is euphemistically called "soft law."[8]

The situation is quite different on the ground. The climate exodus is already here and growing by leaps and bounds with every passing day. Look at two of the hotspots to get a close-up sense of the magnitude of the tragedy unfolding: the millions of climate refugees pouring out of the parched Middle East, risking lives on rickety boats hoping to reach EU shores and their brethren in Central and South America, with whole families and children riding on the backs of their parents, walking for hundreds of miles across treacherous terrain hoping to slip across the U.S. border.

New climate studies are not encouraging. It is estimated that by 2070, "extremely hot zones, like in the Sahara, that now cover less than one percent of the earth's land surface could cover nearly a fifth of the land, potentially placing one of every three people alive outside the climate niche where human beings have thrived for thousands of years."[9] A 2020 study warns that "the geographical position of this temperature niche is projected to shift more over the coming 50 years than it has in 6,000 years," signaling the greatest human climate migration in all of history.[10] Another study, published in the journal *Science*, reported that by 2100 temperatures in parts of the world, including dense populations in India and China, could be so high that going outside even for a few hours "will result in death even for the fittest of humans."[11]

In large swaths of the subtropics and mid-latitudes "the great migration" has already begun – the trickle of climate refugees a mere thirty years ago is now a tsunami. Millions, and within a few decades hundreds of millions of climate refugees, will be scouring the world for safe climate havens. The options are limited . . . either plan ahead or risk utter chaos. If the former, the UN nations will need to bite the bullet and issue formal climate passports accompanied by regulations, codes, standards, and protocols to assure orderly and safe passage. And left hanging is that many nations in the subtropics and mid-latitudes will either wither or cease to exist. We're already seeing this come to pass in the island nations of the Pacific – soon we will see it elsewhere. No amount of head-scratching is going to save the day. There will be nations that endure and others that shrivel and decline or even disappear over the course of the next 150 years, – a short timeline that separates the birth of grandparents from the death of their grandchildren.

How will today's climate migrants know the route? Our climate projections are already pointing the way, and it leads from the mid-latitudes and subtropics to the vast unpopulated lands of the far north – Canada, the Arctic, the Nordic countries, Russia, and Siberia. Much of these virgin land areas, mostly covered by permafrost, now melting, make up 25% of the northern hemisphere's terrestrial surface and can absorb large numbers of our species.[12] Though these northern regions are also quickly warming, our climate scientists tell us that if we can stabilize the planet's temperature at closer to 1.5 degrees Celsius rather than two degrees Celsius by mid-century with a concomitant rewilding of one-third to one-half of the Earth's landmass and oceans over the next several centuries, our species and many of our fellow creatures might survive and even flourish in new ways. But the future is narrowing with each passing decade.

It will be essential to ready climate passports so that nations of moderate to high risk, in the throes of climate change, can downsize their populations and lighten their ecological footprint, rewild their ecological space, and renew their ecosystems. This will allow sizeable numbers of the population to migrate north to less inhabited ecosystems, scattering our species' ecological footprint and lessening our load on the planet – providing a pragmatic approach forward to a challenging future on Earth.

The question confronting climate migrants will be how to map the best environmental route that will allow safe passage to more environmentally compatible host countries which can provide a suitable habitat to sustain large numbers of our species. Scientists are already mapping the routes that climate refugees will need to take to stay aligned with robust ecosystem services on their journeys and layovers on their way to new bioregions that can sustain their numbers. And, it turns out, the guides on the journey are our fellow creatures – mammals, birds, reptiles, amphibians, fish, etc. Like us, our fellow creatures are trapped in a rewilding hydrosphere wreaking havoc on local ecosystems and are fast becoming climate migrants. They too are on the move in search of more hospitable environments that mimic what their biology is accustomed to for survival.

Scientists have begun to track species' migratory routes and are mapping their journeys across wildlife corridors to new destinations that allow them to survive and flourish. A 2018 study published in the journal *Global Change Biology* using high-performance computing methods was the first of its kind that "identified climate connectivity areas across North America by delineating paths between current climate types and their future analogues that avoided non-analogous climates and use centrality metrics to rank the contribution of each location, to facilitating dispersal across the landscape."[13]

While much of the current attention is on saving and restoring the most at-risk ecosystems, the fact is that a rewilding hydrosphere has already moved the climate these species have been accustomed to mostly to the north which makes finding new host regions whose climate is similar to the ones they have left a priority. This doesn't mean giving up on climates that are dramatically warming, but only that the planet's hydrosphere is a much more powerful player in determining how these current at-risk ecosystems will rewild, survive, and flourish. The point emphasized by researchers is that the "persistence of many populations will hinge on their ability to disperse and colonize habitat which has become newly suitable for their climatic requirements."[14]

This study and others to follow will be able to identify the most promising wildlife corridors that species are taking, and allow government agencies and scientists to pinpoint trouble spots along the way, where impediments are blocking the migration routes – be they roads, fences, artificial reservoirs, and other human infrastructure – and clear them away, allowing for safe passage. The mapping of these wildlife corridors includes ensuring suitable ecosystem services to help species speed their journey: for example, reforesting where necessary to provide shade and cover and a cooling effect along the routes. There's also discussion about introducing other features to bolster ecosystem services on these wildlife corridors; features that can enhance a safe passage. For example: introducing beavers, who can establish small dams to connect floodplains and allow water to flow more slowly, will minimize the impact of droughts and be critical to ensuring the well-being of the many species that are migrating along these routes to new habitats.[15]

On March 21, 2023, the executive office of the President of the United States' Council on Environmental Quality issued a memorandum to the heads of all federal departments and agencies on "recognizing the importance of ecological connectivity and wildlife corridors as human development degrades, eliminates, and fragments habitats and as climate change alters environmental conditions." The White House memorandum emphasizes the point that connectivity is:

> the degree to which landscapes, waterscapes, and seascapes allow species to move freely and ecological processes to function unimpeded. Corridors are distinct components of a landscape, waterscape, or seascape that provide connectivity. Corridors have policy relevance because they facilitate movement of species between blocks of habitat, notably during seasonal migration or in response to changing conditions.[16]

Connectivity allows wildlife to access needed resources and facilitates ecological processes. Furthermore, "connectivity promotes climate adaptation and resilience by enabling wildlife to adapt, disperse, and adjust to changes in the quality and distribution of habitats, including climate-driven shifts in species' geographic ranges."[17] The presidential directive calls on all federal agencies to identify "actions that have been or will be taken to advance the objectives set forth in this guidance."[18] What's particularly important about the new scientific studies mapping wildlife corridors and the President of the United States' detailed guidance document is that the United States, as well as other countries, will have a new scientifically expansive topographical approach to mapping the wholesale migration of species, including our own, to new hospitable ecoregions.

The new nomadism will be following the lead of our fellow creatures across migratory corridors, as our ancestors did for 95% of the time our species has inhabited the Earth, but with a caveat. Our path will need to run parallel, but at a distance along the way, so as not to interfere with other species' stopovers and safe journey to new microclimates. To be alongside but at arm's length is the new rule of thumb. To speed the journey, we'll be able to use sophisticated scientific mapping and Internet of Things infrastructure to chart our way, including where to locate temporary pop-up ephemeral cities nearby but at a distance from migratory routes.

These shared journeys across migratory corridors and the settling into more compatible ecosystems, at least for a period of time, is of a different political dynamic. But what we have not yet come to terms with is what these mass exoduses and continuous resettlements leave behind. Entire regions, once under the sovereign governance of nation states, will downsize and be hollowed out as they become unlivable, and even disappear. Regions once under the sovereign governance of nation states will likely cease to be. Much of what we have come to know as civilization, particularly along the subtropics and mid latitudes of the planet, will be abandoned and left to rewild in new ways by a new sovereign – the hydrosphere. If this seems shocking, recall that whole civilizations have come and gone over the past 6,000 years; the victims of both entropic collapse and dramatic changes in the climate. Still, each time, our species dispersed and/or migrated elsewhere and occasionally returned after a period of time to resettle and rebuild anew.

This time around, the timeline is different. The warming of the planet from the emission of global warming gases is taking us at least partially into an extinction event – whether limited in scale or otherwise – and, with it, a new climate regime that will likely be here for millennia to come. Nation states are already hollowing out in the Mediterranean and Central America as a warming climate makes human life uninhabitable. By the end of the twenty-second century, the climate markers suggest that many regions now under sovereign political rule will either wither or perhaps even disappear altogether – that's already a potential reality for hundreds of Pacific Island nations.

Mass migration and resettlement henceforward is likely to be increasingly attached to regional ecosystems, overseen partially as administered commons and in the form of bioregional governance. That transformation has already begun. As mentioned earlier, the first-of-its-kind bioregional governance has been established in the European Union, bringing together the autonomous region of Catalonia and the Balearic Islands in Spain and the region of Occitania in France, establishing the European Pyrenees Mediterranean Bioregion.

In 1991, five Pacific Northwest U.S. states and five Canadian provinces and territories established the Pacific Northwest Economic

Region to share their ecosystems as a governing commons. Between 1983 and 2015, the eight U.S. states and two Canadian provinces that share the Great Lakes ecosystem chartered the Conference of Great Lakes and St. Lawrence Governors and Premiers to steward their bioregion as a commons. Both of these bioregional governances collaborate legislatively to oversee their bioregion and ensure that new business opportunities, employment, and development align with promoting their ecosystems. Nor are these two bioregions anomalies. Rather, they are first movers. In the decades ahead, nation-state governments will become increasingly porous, sharing their common bioregions. Their very survivability will depend on it.

Military Defense Gives Way to Climate Resilience

Let's circle back to the water–energy nexus – a little known phenomenon that is the lifeblood of the Age of Progress, the whole of the industrial revolution, the rise of capitalism, nation-state governance, and the globalization of trade. Keep in mind the earlier mention that power plants use coal, oil, and natural gas to boil large amounts of water to provide steam, which turns the turbines and generates electricity. Recall, in the U.S. alone thermal power generation accounts for a significant percentage of total water withdrawals.

Securing the water–energy nexus lynchpin requires a huge military presence to protect and defend coal deposits, oil and gas fields, oversee dams and reservoirs, provide security for canals and sea routes, and protect pipelines that keep the entire system "operative." Left unsaid is the extraordinary toll on lives in building out these massive infrastructures. Around 120,000 laborers perished from dysentery, smallpox, hepatitis, and tuberculosis during the eleven years it took to excavate and deploy the Suez Canal connecting the Mediterranean Sea and the Indian Ocean, and 30,000 men died from yellow fever and malaria while excavating the Panama Canal to open a waterway between the Atlantic and Pacific Ocean.[19]

European historians remind us that World War II was at least partially precipitated by the longstanding rivalry between Germany and France over control of the rich coal deposits in the Ruhr Valley. Likewise, in July 1941, the U.S. embargoed all petroleum exports to Japan and, shortly thereafter, Great Britain and the Dutch East Indies quickly did the same, choking off Japan's access to oil, without which it would have no way to secure its military expansionism across China, Southeast Asia, and the Pacific, or even keep its own economy up and running.[20] Four months after the oil embargo, Japan bombed Pearl Harbor and entered World War II. Seventy to eighty-five million people perished in the war, or about three percent of the total population of the world at the time.[21] It's not a stretch to say that millions of human beings have gone to their deaths in building out the water–energy infrastructure and defending fossil fuel assets over the course of the past two centuries while working to secure the Industrial Age and make good on the promise of advancing the Age of Progress.

Now, the nexus of hydraulic and fossil fuel infrastructure is taking society along two divergent paths. Global warming from the burning of fossil fuels has tipped the planet's hydrological cycle into an accelerated feedback that is changing the way that water is distributed across the Earth, impacting ecosystems and human infrastructure in ways that are inimical to how human civilization is modeled and managed. At the same time, as already mentioned, the levelized cost of utility scale solar and wind-generated power in 2019 dropped below the cost of all other conventional fuels – uranium, coal, oil, and natural gas, and is continuing to plummet, opening up the opportunity of quickly moving beyond the fossil fuel era. The market is speaking.

Yet, the world's military superpowers – the United States, the European Union, Russia, and China – and for that matter militaries everywhere, continue to base their security on protecting and defending the traditional water–energy nexus infrastructure of a fossil fuel-based industrial civilization, despite its irreversible exodus from the world stage. In the past few years, however, the realization has begun to set in that the fossil fuel water–energy nexus has run out of steam and a new narrative is coming to the fore, taking us from the Age of Progress to the Age of Resilience, and from a highly centralized and vertically integrated water–energy infrastructure to a highly distributed water–energy platform made up of local and regional water catchments and the energy of the sun and the wind that are now cheaper and safer than nuclear, coal, oil, and natural gas. And, unlike fossil fuels and uranium, which are only found at scale in a few geographic pockets and are costly to find, extract, refine, transport, consume, and dispose of, the sun and the wind are everywhere and are now cheaper than all of the conventional fuels of the Industrial Age.

As already mentioned, millions of homeowners and hundreds of thousands of businesses, neighborhood associations and urban and rural communities are harvesting sun and wind electricity where they work and live. The surplus power they are not using is being sent back on an increasingly digitized electricity internet to be shared across regions, time zones, and continents, and within the next decade across oceans, allowing billions of human beings to share renewable energy. The transformation from the Age of Progress to the Age of Resilience comes with a total rethinking of the nature of security from capturing and enclosing the hydrological cycle and extracting and protecting fossil fuels to adapting to a rewilding hydrosphere along with the appropriation and sharing of solar and wind-generated electricity.

This reality check is now at the front door and the subject of increasing urgency in militaries around the world. On October 10, 2018, the Tyndall Air Force Base in the Florida Panhandle was devastated by 150-mile-an-hour winds from Hurricane Michael, a category five hurricane. Ninety-five percent of the buildings were severely damaged or destroyed. Tyndall is the home base of the Air Force's F-22 Raptor Stealth Fighters that were severely damaged in the hurricane. The carnage and financial losses at this single airbase exceeded all the losses experienced by the United States from Iranian missile strikes in the Middle East.[22]

A 2019 Pentagon report concluded that 79 domestic U.S. military bases are currently vulnerable to drought, water scarcity, wildfires, floods, and hurricanes, in addition to more than 1,700 U.S. global military installations along coastlines that are potentially vulnerable to rises in sea level.[23] Of course, the irony of all this is that the U.S. Department of Defense is the single largest institutional generator of global warming gases on Earth.[24] Between 2001 and 2017, the U.S. military generated an estimated 766 million metric tons of CO_2 emissions. That's more global warming emissions than many smaller nations.[25] This situation is compounded by the fact that the U.S. Department of Defense and the military establishments of other superpowers are defending a fossil fuel-based civilization that is quickly sunsetting, and whose stranded assets – exploration rights, oil rigs, pipelines, refineries, storage facilities, building stocks, power lines, telecommunications grids, logistics infrastructure, dams and artificial reservoirs, etc. – are estimated to run into the tens of trillions of dollars.[26]

In June 2021, the International Military Council on Climate and

Security surveyed the world's leading institutions who address future risks to society, and found nearly unanimous agreement that the question of security is beginning to shift from defense of territory to rescue and recovery of populations, the restoration of civic order, and the rebuilding of infrastructure brought on by climate-related disruptions and disasters. The report concluded that "respondents expect a majority of [climate] risks will pose high to catastrophic levels of risks to security. Ten and twenty years from now, respondents expect very high levels of risk along nearly every type of climate security phenomenon."[27]

In the fall of 2021, the U.S. Department of Defense signaled a possible shift in its own thinking in regard to mission priorities, with the issuing of two seminal reports. In one of the reports, the DoD laid out a risk security analysis whose findings were daunting. The report concluded:

> as the frequency and intensity of these [climate disasters] increase, impacts are likely to expand competition over regions and resources, affect the demands on and functionality of military operations, and increase the number and severity of humanitarian crises, at times threatening stability and security . . . however, in worst-case scenarios, climate change-related impacts could stress economic and social conditions that contribute to mass migration events or political crises, civic unrest, shifts in the regional balance of power, or even state failure. This may affect U.S. national interests directly or indirectly, and U.S. allies or partners may request U.S. assistance.[28]

In the second report, the U.S. Department of Defense issued a long-anticipated climate adaptation plan and at last zeroed in on a partial change in mission by "identifying climate-impacted problems early" in order to "better prepare for humanitarian assistance and disaster relief" and "adapt or strengthen military engagement programs to prepare partners to face potential climate-sourced conflicts." Unfortunately, the DoD rationalized the partial change in mission in mostly geopolitical terms than biosphere terms by suggesting that it will "give the DoD an advantage over adversaries and competitors to better survive and operate in future environments."[29]

The U.S. Department of Defense and militaries in every country are totally unprepared to meet the challenge of a transformational change in the planet's hydrological cycle brought on by climate change and, while engaging in rescue, recovery, and rebuilding on a case-to-case basis, continue to think of national security almost exclusively in geopolitical terms. The bottom line is that militaries are at best thinking of climate risks more narrowly in regard to their own installations and how to build resilience against climate disasters that might cripple their operations, but have failed to seize on the bigger picture of the change in mission itself – that is, the need to shift from geopolitical security to biosphere resilience by assisting populations in preparing, surviving, recovering, and adapting to the planet's quick-evolving hydrological cycle.

If necessity is the mother of invention, however, the responsibility of addressing the escalating carnage to ecosystems and society from climate change and the transitioning of the hydrological cycle is, by default, being heaped on the shoulders of the U.S. military and militaries in every country. In the U.S., federal troops and states' national guards are increasingly being deployed on the domestic front as first responders: fighting wildfires, rescuing populations from floods and hurricanes, providing water, food, medicines, and temporary lodging, and even rebuilding infrastructure. In 2017 alone, to take one year, the hurricane season in the U.S. – one of the deadliest in history – required a huge military rescue and recovery and "involved 43 states and resulted in the activation of more than 50,000 members of the national guard."[30]

The long-established agencies of the U.S. Department of Defense might take a cue from the newest branch of the U.S. armed forces, the Space Force, which was established in 2019 with the mission of organizing, training, and equipping space forces in order to protect the U.S. and allied interests and provide space capabilities to the joint force. As part of its mission, however, the Space Force is at least partially focusing on monitoring climate hotspots with the goal of anticipating climate disaster events and sharing the data with other countries, allowing them to prepare for climate resilience and not just rescue and restore.[31]

PART IV

Sublime Waters and a New Ontology of Life on Earth

CHAPTER 12

Two Ways to Listen to the Waters

Learning to Live on Planet Aqua is an exegesis on rethinking humanity's long-tangled relationship to the hydrosphere. In 1751, Edmund Burke published a work entitled *A Philosophical Enquiry into the Origin of Our Ideas of the Sublime and Beautiful,* which had an ontological impact on both the European Enlightenment and the Romantic Period that followed, and subsequent cultural narratives embedded in the Age of Progress till present times. Today, with the specter of climate change menacing our every moment in real time, the notion of the sublime has re-emerged in academic circles, but this time the stakes are much higher, with an extinction event looming on the horizon. At the core of the calamity befalling the Earth and all forms of life thereon is the hydrological cycle bursting forth in unexpected ways.

Burke describes the feeling of vulnerability our species feels when suddenly confronted by the overwhelming presence of nature's agencies, be it in a passive mode like the great mountain ranges looming over the land or deep canyons sculpted by the waters, or more often an active mode including powerful waterfalls, sudden avalanches, terrifying hurricanes, torrential floods, incendiary wildfires caused by droughts and heatwaves, spiraling tornados, volcanic eruptions, shattering earthquakes and unexpected tsunamis, etc. These up-close experiences can cause terror, but if experienced from a safe distance can evoke the feeling of pure awe confronted by the power and majesty of the Earth's agencies. That feeling of awe often precipitates a sense of wonder and sparks the human imagination. It's at this juncture that imagination can follow along two very different paths.

The first is sketched by the philosopher Immanuel Kant, who suggested that while such sublime experiences trigger not only a sense of helplessness but also awe, wonder, and imagination at the power of nature, the rational mind – a nonmaterial force immune from the tempest of the physical world and more powerful than even the forces of nature – steps forward with its innate sense of reason and takes a measure of the sublime experience, captures it, and tames it to the practicalities of enhancing human presence on Earth.

Others, like Kant's nemesis, the philosopher Arthur Schopenhauer, suggested a different and opposite reaction to the sublime experience. He agreed with Kant that a sublime moment draws an immediate reaction of cringing and even cowering in the face of nature's force, leaving the observer momentarily numb. But, upon regaining one's wits and sensing the awe and wonder of nature's presence, Schopenhauer imagines a radically different human response to the sublime that takes our species on a divergent journey in its relationship to our earthly abode.

Schopenhauer believed that the sense of awe, wonder, and imagination sparked by the sublime is as likely to unleash a feeling of the majesty of nature and lead to an emotional and compassionate embrace – to wit, a transcendent moment of identification with the natural world to which we and our evolutionary relatives are beholden as fellow travelers on a life-affirming planet. Musing on the feelings welling up by the sublime, Schopenhauer was of two minds. On the one hand, when confronted by the overpowering presence of nature's forces, the individual feels utterly small and insignificant and of little or no account. He writes, "he feels himself as individual, as the feeble phenomenon of will, which the slightest touch of these forces can annihilate, helpless against powerful nature, dependent, abandoned to chance, a vanishing nothing in face of stupendous forces." On the other hand, "he also feels himself as the eternal, serene subject of knowing, who as the condition of every object is the supporter of this whole world . . . This is the full impression of the sublime."[1] For Schopenhauer, the feeling of the sublime kindles one's sense of intimate belonging to the larger forces that animate life, comforted by the thought that we are each at one with a vibrant Earth.

The great debate between Kant and Schopenhauer some two centuries ago over the nature of the sublime and the human response is more timely now than ever before, as our collective humanity wrestles with two different approaches to the mass extinction of life on Earth that now lies before us. The waters are roiling in reaction to the warming of the Earth from global heating gases, taking the hydrological cycle of the planet on a wild ride. Our species' future depends on how we choose to relate to the waters that animate life.

Kant would have us exercise our rational impulse and force the waters to adapt to our species' whims while Schopenhauer would have us empathize with the life-affirming nature of existence and find ways to adapt to a fast-changing hydrological cycle. Humanity will need to choose between these two very different approaches to the future on the Blue Marble. The decisions we make of which narrative to pursue will affect not only our destiny but the future of life itself on Earth.

Vanquishing the Waters or Riding the Waves

Listening to the waters and feeling at home immersed in a water planet will not be an easy transformation, especially in the Western world, where we have perceived space as place and place as solid ground upon which our spindly bipedal frame is firmly ensconced. Although it's true that we are among the mammalian species who can tread water and even voyage across great oceanic expanses, we think of ourselves as of the Earth. Our cousins, the whales, have proven to be more versatile. They began as small deer-like land-based mammals and, for some unknown reason way back in the evolutionary record, made the decision to leave the land and, over time, repopulate in the oceans as whales.[2] Less versatile, our species stayed on the land and came to think of the waters as the "great other," at least in the Western world.

At a deeper cultural level, the Western mindset has been groomed to think of the waters as an alien force, a dark void, a theater of chaos. Our primal fear of the waters, in large part, lies buried in our unconscious and is rarely examined, except in those times where we are stranded on the oceans and surrounded by the deep, or caught up in a fierce hurricane or tsunami, suddenly finding ourselves helpless and adrift and terrified by the sense of being swallowed up by the void and lost forever.

Interestingly, with the dawn of the hydraulic civilizations and the great axial religions that accompanied them, the Western and Eastern civilizations parted ways in their relationship to the waters.

Although both the West and Asian geographies shared a passion for controlling and directing the planet's hydrosphere – adapting the waters to our species' cravings and whims – the Western axial religions, Judaism, and later Christianity and Islam differed from the Asian axial religions in how they perceived the waters: an ontological schism that would come to define the very different cultural evolutions of the West and the Eastern psyches. The Abrahamic account of the Creation in the opening pages of the book of Genesis, as mentioned earlier, begins with the waters that preceded Creation. Jehovah willed there to be light, separating day from night, and created the land atop the waters, followed by the creation of the Garden of Eden and the sculpting of Adam from the dust of the Earth. The Abrahamic axial religions prioritized the land over the waters, regarding the latter as the abyss, and set their sights on being swept up into heaven in the next world.

In the first verse of the twenty-first chapter of the Book of Revelation of St. John the Divine, the prophet describes the coming of a new Garden of Eden after the apocalypse. "And I saw a new Heaven and a new Earth: for the first Heaven and the first Earth were passed away and there was no more sea."[3] The vanquishing of the sea was to be the final victory of light against dark and a guarantee of order prevailing over chaos and the triumph of everlasting life.

Lest we put too much responsibility on the Abrahamic vision of an epic struggle between the heavens, earth, and the dark deep, suffice it to point out that similar creation stories were told as far back as the first hydraulic civilization in ancient Mesopotamia. Mircea Eliade points out that the taming of the waters across cultures was conceived of as a great struggle between chaos and order, the void and structure, in the rise of every hydraulic civilization. The Abrahamic tradition, however, was unique in making the struggle between *terra firma* vs. aqua the centerpiece of its worldview. Christopher Connery, an historian at the University of California at Santa Cruz, observes that in this respect, the West and Asian hydraulic civilizations parted ways. He writes that in China:

> rarely did the ocean have the same status, figuratively and cosmogonically, that it did in the West. Imperial Chinese discourse

> on the ocean, whether during the time of its maritime dominance, or during its retreat from maritime activity in the fifteenth century, contained nothing like the personified or abstract character of the ocean as elemental power that was so common in the West . . . At no point in imperial history did Chinese rule appeal to the hydraulic function as its *raison d'être*: flood control was one of government's duties, but not the defining function. Imperial China had no tradition of ocean-voyaging literature, no significant body of poetry on the sea, and despite having one of the most developed and most ancient traditions of landscape painting, had no tradition of seascapes. In short: China, one of the pre-modern world's great maritime powers, had no strong tradition of oceanic figuration.[4]

The religions and philosophies of Daoism, Buddhism, Hinduism, Shintoism, Confucianism, etc. were wedded to a very different relationship to nature. They rejected the concept that human beings are privileged over the whole of the natural world. Although when it comes to beliefs versus practices, China, India, Korea, and Japan, etc., clearly exercised a very similar approach to yoking the waters, forcing them to adapt to our species' agency than the other way around. Still, they did so with mixed feelings. The concept of harmonizing with nature is so entwined in Eastern religions and philosophical traditions that it tempers the Western theological fervor to dominate and vanquish the deep. While Asian cultures often stray in practice, they never see the waters as the great void that must be vanquished. China, India, Korea, Japan, and much of Asia will likely have an edge in making the turn from the Age of Progress to the Age of Resilience and from geo-politics to biosphere politics as they seek a harmonious reconciliation with the hydrosphere.

At the 18th National Congress of the Chinese Communist Party in 2012, a rare amendment was made to the party's constitution, formalizing a change in the prioritization of its development by committing the nation to a strict alignment of all economic and social development to the requisites of nature – the first such country in the world to do so, opening up the next great era in our species history. At the 19th National Congress of the Communist Party in 2017 the delegates revised and strengthened the country's commitment to ushering in what it called an "Ecological Civilization."

In 2019, President Xi Jinping wrote and published a landmark document entitled *Pushing China's Development of an Ecological Civilization to a New Stage* – a sweeping new vision calling upon China to lead the world in an historic re-enchantment with nature – that is seeking harmony and adapting our species to nature rather than forcing nature to adapt to our narrow and limited approach to commodifying, propertizing, and consuming the planet's plenitude for short-term gain but at the expense of despoiling the Earth. Xi's philosophical essay marks a fundamental cultural difference from the traditional Western approach of asserting mastership and dominion. President Xi Jinping wrote:

> Throughout the process of development, we must prioritize environmental conservation and protection and put the restoration of nature first ... however, we must be soberly aware that in our efforts to build an ecological civilization, we face grave challenges, immense pressure, and prominent contradictions. The situation remains extremely serious ... there are still many difficult obstacles that we must overcome, tough problems that we must face, and longstanding issues that we must resolve ... protecting the environment is a challenge and a duty which all of us around the globe must face together.[5]

In 2022, China announced "A Climate Adaptive and Resilient Society by 2035." This initiative, led by the Ministry of Ecology and Environment, with the participation of sixteen other national ministries, offers a systemic shift in the economic and social development priorities of the country. The new plan moving forward elevates "adaptation to climate change" along with its existing commitment to "mitigating global warming emissions," recognizing that both narratives combined provide the grist for establishing an ecological civilization.[6]

The Age of Resilience plan is sweeping and thoughtful, introducing a whole new venue of economic, social, and political practices, including the introduction of new levels of governance with the formulation of eight ecoregions – joining the ranks of the U.S. and Canada governing bioregions. China is also among the handful of nations to introduce a new approach to scientific inquiry in the form of complex adaptive social/ecological systems modeling, tailored to an ecological civilization where the culture/nature divide is severed, and our species begins to think of itself as only one of multitude of species and earth agencies that make up a self-organizing and ever-evolving planet.

China's official embrace of establishing an ecological civilization requiring society to adapt to nature rather than nature adapting to society, is a bellwether of the second great reset of our species relationship to nature and, hopefully, a vision that nations, regions, and localities across the world will formally adopt in their own constitutions, laws, and deployment in the decades ahead.

Adapting to nature and letting the waters run free seems so appealing at first that many would say, "why not?" without a second thought as to the full implications of such a prospect. The romance of it is alluring. But, those waters are nevertheless going to find their own way back onto the old floodplains, and the long hidden away streams and rivers are going to bubble up anew across urban and suburban communities that were once their domains. It's no longer our choice to make. The hydrosphere is now dictating the terms regarding the future evolution of the planet – the intensification of the snows, floods, droughts, heatwaves, firestorms, and hurricanes is already forcing our hand. Hundreds of millions and likely several billion of our human family will be forced to migrate to safer havens where they can find livable habitats and adapt to the waters.

If we really want to rid ourselves of preconceived notions about our ability to sequester the planet's four powerful spheres, consider the wildfires that are blazing across the lithosphere by the rewilding hydrosphere. David von Drehle, a *Washington Post* columnist, penned a provocative column in the early fall of 2022 entitled "How to Prevent Deadly Wildfires? Stop Fighting Wildfires" that likely caught readers off guard. He drew his readers' attention to a book written a quarter of a century ago by Norman McLean entitled *Young Men and Fire*, in which the author retold an incident that occurred on August 5, 1949, when fifteen U.S. Forest Service elite airborne fighters – smoke jumpers – dived down into a remote Montana gulch of dense pine tree forests and waving fields of grass. A sudden change in the direction of the wind hours later swept over the firefighters, leaving twelve of the fifteen men consumed by the blaze.[7]

A tough lesson was learned from this tragedy that speaks to the bigger issue of whether we should let some of the wildfires go so they can run their course – at least if they are not a threat to population centers. Even then, if entire communities are nestled nearby or within dense forest areas, wouldn't it be better to plan ahead and make peace with the inevitable by relocating entire communities to safer ground instead of trying to win a battle that is unwinnable?

Von Drehle notes that most of us are unwilling to speak truth to the limitations of human agency, saying that:

> for more than a century, we've pursued the idea that the only good fire is an extinguished fire, and along the way our wilderness has filled with flammable fuels. Fires get bigger, and move faster, when there is more to burn . . . Fire is [oftentimes] nature's way of renewing wilderness by clearing excess growth and dead fuels.

By letting nature control the fires, or at least welcoming what firefighters call "controlled burns," society limits the extent of the losses but also assists nature in its own cleansing. We have to become comfortable with the fact that "artificially thwarting the necessary work of wildfire only makes future fires worse by giving them more fuel."[8]

While in theory most of us might think this likely makes sense, in daily life we are so used to controlling nature and forcing it to adapt to our conveniences and lifestyles – for example, living on paved-over floodplains or building new communities surrounded by ancient forests, or establishing urban centers in arid deserts to satisfy our aesthetic desires, we pay little heed to the ecosystem dynamics of our surrounding hydrosphere, lithosphere, atmosphere, and biosphere.

The luxury of adapting nature to our species is forever over. Climate change and a rewilding hydrosphere now hold all the cards. The question is – will we awake from our long slumber and begin anew by learning how to adapt our species to a rewilding of the hydrosphere and its sister spheres that are fast changing the ecological dynamics of life on Earth?

Decommissioning the dams and freeing the streams and rivers to take their natural course, letting the oceans take coastal regions down under, and migrating to new more hospitable terrains is already beginning to happen and will increase exponentially in the decades and centuries ahead. Getting ahead of the curve and preparing for a new semi-nomadic and ephemeral way of life is inevitable. Our species is blessed with the cognitive ability to assess what's coming and the empathic sociability wired into our neurocircuitry to band together and cooperate as a species and, lastly, the biophilic sensibility to act on behalf of the interest of our fellow creatures – our extended evolutionary family – to make the leap and reset our journey. Relearning how to adapt to nature's calling using all of the sophisticated mental acumen at our disposal is the signature of an emerging Age of Resilience that's leaving a worn Age of Progress behind.

Reorienting Ourselves in Time and Space in a Liquid Milieu

It was the ancient Greek philosopher Heraclitus who remarked "no man ever steps into the same river twice." A *terra firma* bias favors being over becoming, while an aqua orientation favors the opposite – two very different ways of distinguishing the meaning of time, space, and existence.

Our very concepts of time and space, which lie at the heart of all philosophical debate and shape the way that we have come to think of reality, are partially conditioned by whether we embrace an earth-centric or aqua-centric way of thinking. The earth-centric approach during the whole of the industrial era, and even earlier, has been to conceive of earth as containing passive resources in the form of distinct objects and structures that can be easily commodified, propertized, exploited, and consumed in a timeless vacuum. The aqua-centered approach sees the waters as processes, patterns, and flows. Water doesn't exist at an instant and it can't be easily pacified, held down, and secured. What's incontestable is that the waters are far and away the most powerful agency on the planet but, in the long history of hydraulic civilization, our species has lived with the mass delusion that the whole of the waters on the Blue Planet could be tamed and domesticated – a naivete that time and again has led to the inevitable dissipation of entire hydraulic empires, all brought down, in large part, by the entropic bill that inevitably trails alongside the enclosure of the waters.

If there was a defining moment in our species' relationship to time, space, and the waters that gave rise to the industrial era and the globalization of commerce and trade, it would be two inventions that came along around the same time in history. Recall that in the late fifteenth century Spain and Portugal entered into a treaty by which they divided up the world's oceans between them, with each claiming sovereignty over roughly half of the oceans of the planet. While their agreement was audacious, unlike the land, the oceans are more fluid, ever flowing, always expanding and contracting, and often changing course, with their currents complicated to navigate, much less control, all of which made it near impossible to sequester, master, and privatize. Early ocean wayfarers were roaming the seas in search of new lands, but with little more direction than relying on legends, hearsay, and subjective accounts by earlier expeditions that were unreliable. That all changed with two technological advances that would come to alter the course of history and lead to the propertization of vast stretches of the oceans that constitute the Blue Marble.

The first invention was the Gutenberg print press, circa 1440s, which transformed the very way our species communicates and, in the case of seafaring, changed the way ocean explorers organized space. With print, new discoveries of coastline demarcations could be compared, synthesized, standardized, and printed out, eliminating subjective bias and inaccurate descriptions, allowing seafarers far more accurate navigational maps.

The standardization of maps greatly improved ocean travel and the exploration of new continents. Yet, there remained the problem of not being able to calculate longitude on the high seas. The issue of longitude was about calculating time. When Columbus set sail across the Atlantic in 1492, he was able to measure latitude from observations of the North Star, but there was still the problem of measuring longitude, without which sea travel and the opening of the new world for exploration and enclosure would remain a quixotic enterprise, and printed maps would have limited value.

The inability to measure longitude meant ships were not able to ascertain the most direct route, or a route with the best winds and currents, slowing voyages for days and even weeks. The time delays often imperiled the crews, with rations running out. Errors in reckoning position also caused deadly shipwrecks at sea. To find a solution to the problem, the British government established the Board of Longitude in 1714 and offered a prize for anyone who could determine it. The solution was to carry a mechanical timepiece on board the ship, which could maintain the correct time at a reference location. The problem was that there was no clock capable of keeping accurate time on rolling and pitching ships. Finally, one John Harrison, after much trial and error, succeeded in constructing a marine chronometer and was awarded a handsome financial prize in 1773 for his feat.

The coming together of standardized printed maps and a working chronometer to calculate longitude on the waters opened oceanic exploration, the colonization of the continents of the new world, global trade, and the systemic enclosure of the oceans and landmasses. Today, a new technology revolution is once again changing the way we measure and navigate time and space and it's the Janus face of the motivation and drive that compelled earlier explorers to traverse the oceans, seize continental land masses, and propertize the waters. The new technology is called the Gravity Recovery and Climate Experiment (GRACE). NASA, the Jet Propulsion Laboratory, and the German Aerospace Center launched two satellites in 2002 to measure the Earth's gravity fields for anomalies. The GRACE mission is to record gravity anomalies to pinpoint how the Earth's mass is distributed along the planet's surface and how it changes over time. The monthly gravity anomaly maps are 1,000 times more accurate than any previous maps, allowing hydrologists, oceanographers, glaciologists, and climate scientists to study the impacts that climate change is having on the hydrosphere, lithosphere, atmosphere, and biosphere in real time. The datapoints keep a running tab on the thinning of the ice sheets, the flow of water in aquifers, drought conditions, and even the currents of magma embedded in the Earth's core. The mapping even includes ocean bottom pressure and detecting changes in deep ocean currents.

GRACE mapping provides data in real time on sea-level rise and whether it is being caused by melting glaciers, a thermal expansion of warming waters, or even changes in the salinity of the ocean. The GRACE categorizing of the waters is so fine-tuned that it can detect exactly how much ice is lost per year in places like Greenland and Antarctica and how much water is depleted in real time in India. The GRACE cartography even provides stacking data. Using GRACE tracking and data, the University of California at Irvine published findings between 2003 and 2013 on 37 of the planet's largest aquifers, and found that 21 "have exceeded sustainability tipping points and are being depleted."[9] The data GRACE is collecting are informative, giving us invaluable insights into how climate change is affecting the hydrological cycle and what is happening on every square inch of the planet, so we can anticipate, prepare, and adapt to the fast-changing waters circling the Earth.

A comprehensive ten-year study of the GRACE data by the Global Institute for Water Security at the University of Saskatchewan was telling. In broad terms, the data point out that the high latitudes, including the U.S., and the low latitudes are both getting wetter, and the mid latitudes in between, which have historically been semi-arid or arid, are getting drier, confirming the collective data in reams of scientific reports issued since 1990 by the UN Intergovernmental Panel on Climate Change. However, unlike the IPCC studies that projected that these changes would become more apparent near the end of the twenty-first century, the GRACE data show that changes are occurring right now. Still more egregious, the data also show that by 2025 over 50% of the world's population will live in water-stressed areas.

Other studies disclose that temperatures are rising faster in the Arctic and Antarctic regions, with a history-changing meltdown of the remaining glaciers of the last Ice Age from 11,000 years ago.[10] The cartographic mapping of the temporal and spatial changes in the Earth's hydrological cycle and their impacts in real time and over time allow scientists, governments, industry, and civil society to anticipate, prepare, and adapt to ongoing climate disasters. By implementing short and long-term climate resilience plans both onsite and across migratory corridors, and even including resettlement and other initiatives, it will allow our species to stay up to the moment in preparedness and be able to secure communities and hopefully thrive in the Age of Resilience.

The GRACE data, analytics, and algorithms represent a technological reset of our species' experience of time and space. The monthly mapping of how the warming climate is causing unparalleled changes in the geomorphology of our planet's hydrosphere is reorienting the way we think about history on Earth and the way we store our calendrical memories and, particularly, momentous climactic events that not only reshape our planet but also our adaptation to it.

Recall, for forager-hunters, the important dates to commemorate were the winter and summer solstice, marking the celebration of the shortest and longest days of the year. The winter solstice marks the ending of the old year and the beginning of the new year and the cyclical rebirth of life. Neolithic monuments like Stonehenge in England, Newgrange in Ireland, and Maeshowe in Scotland were thought to be monuments for capturing the sun on the shortest day of the year to ensure it would light up the world again with the rebirth of life in the new yearly cycle.[11] Neolithic peoples – the Slavic, German, and Celtic societies – would also celebrate the summer solstice in chants and dance rituals along with the lighting of bonfires, hoping "to boost the sun's strength for the remainder of the crop season and ensure a healthy harvest."[12]

The advent of the great hydraulic empires in the centuries preceding the Christian era marked the beginning of "civilization" but also a partial transition from ritual commemorations of the changing seasons to marking the events calendrically in pictographs and later cuneiform. Sumer, the first known hydraulic civilization in the Middle East located on the banks of the Tigris and Euphrates Rivers, not only sported the first urban settlements and written language, but also nurtured the first proto-scientists in the form of palace priests who studied the changing constellations in the night sky throughout the year to better determine the timing of the spring floods so crucial to the workings of hydraulic civilizations. The very viability of these societies depended on accurately predicting the changes in the water cycle over a year in order to organize the many stages involved in managing the hydraulic infrastructure in these first urban cultures.

The common denominator in the calendrical structuring of time and, by extension, space, across all cultures is that they are either marked off by the yearly journey of the Earth around the sun (solar calendars), or the monthly cycles of the moon (lunar calendars), or a combination of the two – called solar/lunar calendars. But, without exception, these solar and lunar calendars have also been frames to fashion the religious, political, and economic narratives of governing cultures.

For example, pictograph calendars included historic events often associated with the exploits of sky gods. Other hydraulic civilizations across the world operated in a like fashion and included similar calendars. The coming together of the great axial religions and their attachment to calendrical cultures has remained the central authority for organizing economic activity, social life, and governance throughout most of recorded history. Each of the axial religions marked off special holy days for repentance, thanksgiving, religious confirmation, and service, which allowed the populace of dense hydraulic social organisms to adhere to a common understanding of the world around them and what was expected of them in time and space. In this regard, calendars are often notorious for rearranging historical moments and dates to further the political ends of their governing regimes.

Before the Christian calendar came into being, the period between December 25 and January 6 marked the celebration of winter solstice rites. The Church was determined to confront these pagan rituals by superimposing the nativity and the epiphany on the same days.[13] Likewise, Church fathers were chagrined by the fact that Easter, the holiest day of the Christian calendar, marking the resurrection of Christ, occurred on Passover, the holiest of Jewish holidays, which is celebrated during the full moon. Worried that the close affinity of these two religious events would undermine efforts to establish Christianity as a separate religion, Church officials met at an ecumenical council in Nicea in 325 and passed a resolution stating that Easter should be observed "on the Sunday following the full moon which coincides with, or falls after, the vernal equinox." This calendrical reform assures that the two holy days would never coincide. The Roman Emperor Constantine issued a letter to the Christian community after the calendrical reform was passed, remarking that "it appeared an unworthy thing that in the celebration of this most holy feast, we should follow the practice of the Jews . . . for we have it in our power, if we abandon their custom, to prolong the due observance of this ordinance to future ages . . . let us then have nothing in common with the detestable Jewish crowd."[14]

Often taken for granted, calendars throughout the history of civilization are the centerpieces around which society is organized. They establish new ways of reimagining the collective memory to fit the new ideas, norms, behavior, and practices governing society. Eviatar Zerubavel, a professor of sociology at Rutgers University, notes that "the calendar helps to solidify in-group sentiments and thus constitutes a powerful basis for mechanical solidarity with the group. At the same time, it also contributes to the establishment of intergroup boundaries that distinguish as well as separate, group members from outsiders."[15]

Not infrequently, new calendrical reforms backfire. For example, the architects of the French Revolution were committed to ridding Western civilization of what they considered to be religious superstition, cruelty, ignorance, and oppression of the Church and State reign of the previous era. They entertained a new image of the future in which reason would reign as a cardinal virtue and usher in a utopian vision of the future. Toward that end, the national convention of revolutionary France on November 24, 1793, put into place a radical new calendrical system reflecting the scientific principles of the French Enlightenment.

Their new calendar de-Christianized time to eliminate Church influence over the French people. The calendar began with changing the

Christian version of the two eras of all of history, which were demarcated by "Before Christ" (BC) and "Anno Domini," or the Year of Our Lord (AD), and placing the birth of Christ in year one. The new French calendar erased these demarcations and established year one as 1792, the birth date of the French Republic, in line with their desire to have a calendar that was both rational and scientific and compatible with the new spirit of the age.

Anxious to orient the French public to a more rational and scientific approach to life commensurate with a nascent Age of Progress, the Republican government introduced a decimal system composed of twelve months, each containing thirty days. Each of the months was divided into ten-day cycles called decades, and each day was further divided into ten hours, and each hour was further divided into 100 decimal minutes, and each minute into 100 decimal seconds, all aligned with the newly devised metric system of measurement.[16] The French Republic also formally introduced the metric system in 1799 as a more scientific and practical means of managing commerce and trade.

If that weren't enough to raise the ire of the French public, calendrical reformers eliminated all of the Christian calendrical holidays, including 52 Sundays, 90 rest days, and 38 holidays, leaving the French people with a tiring work schedule with little reprieve. In their place, the calendar included a handful of rest days with titles like "Human Race Day," "Freedom and Equality Day," "French People Day," "Benefactors of Mankind Day," "Patriotism Day," "Justice Day," "Republic Day," "Conjugal Fidelity Day," "Filial Affection Day," and "Friendship Day." All in all, the new calendar reduced the number of days off to 36, which enraged the French public. The new calendar lasted only thirteen years. In 1806, Napoleon reinstated the Gregorian calendar to appease the French public and placate the Pope.[17]

The influence of the calendar penetrates even further into the unconscious and is unknowingly one of the overlaying instruments for establishing the organization of a people. Political scientist Tom Darby observes that the French Republic calendar served as "a subtle method designed to eradicate the popular consciousness of all previous associations, loyalties, and habits and to substitute in their stead new ones that emphasized the revolutionary ideology. This was to have the double effect of first instituting a state of 'mass forgetfulness' and second of inaugurating the founding of a new popular memory."[18] The same can be said of calendars throughout history.

While the French Republic's calendrical experiment was quickly abandoned, in ensuing decades of the nineteenth century new commemoration days began to appear in stride with the industrial revolution, capitalist markets, and the rise of nation-state governance. The new calendrical markers increasingly paid homage to famous battles, political martyrs, technical discoveries and, most importantly, the establishment of commercial holidays. The few religious holidays that remained in the calendar made sure that the public could hold on to their religious past while adapting to the new commercial age. The calendar only included a paltry number of references to the natural world and environmental markers, clearly indicating the prioritization of ideological and commercial associations over natural and environmental affiliations in how we identify ourselves over time and through space and conceive of the meaning of our existence on the planet.

But if history, at least on Earth, is about "lived" experience, how is it we know so little about the hydrosphere that animates life – this very special gift that may be present only on our planet? And, if our scientists are right and the hydrosphere and accompanying lithosphere, atmosphere, and biosphere are the prime agencies from which all of life emanates, why is it we draw a blank when trying to imagine their temporal and spatial history? Introducing the history of the waters and companion spheres into our calendrical record, including important historical benchmarks, realignments, and even their sequestration and entrainment at the hands of our species would be a welcome first step in rethinking our journey on Planet Aqua.

It's only in recent decades, with the onset of global warming and the dramatic rewilding of the planetary hydrosphere, that scientists across the academic disciplines have engaged in a deep probe of the history and anthropology of the hydrosphere. (Modern global record keeping of the climate only began in 1880.) What the researchers are discovering is fascinating and a first step in understanding the water planet we inhabit. The new hydro data will be important in establishing wholly new calendrical markers, to which we will need to adapt and align if we are to relearn how to live and thrive on Planet Aqua.

For example, scientists have discovered the actual life cycle in time and space of each water molecule, yet only a handful of scientists, technicians, environmentalists, and scattered intellectuals know about it. To begin with, there are over 1.5 sextillion molecules in a drop of water. There are also five sextillion atoms per droplet.[19] With very few exceptions, most of the water on the planet has been here since very early on in the Earth's timeline. A single water molecule is impressive – all the more so because its journey animates both the life cycle of the lithosphere and changes in the oxygen of the atmosphere, while safeguarding the viability of the biosphere.

Scientists use two terms to describe how long a water molecule stays in a given system – residence time and transit time. A water molecule stays in the atmosphere for around nine days and when it descends to Earth it will stay on the ground for a month or two. Some of the water molecules will trickle down into shallow ground water and take up residence for upwards of 200 to 300 years. A deep ground water molecule can remain in residence for up to 10,000 years. A water molecule that comes back from the atmosphere to the ground as snow can remain in residence for two to six months until the spring melt, whereas a molecule that descends from the atmosphere to the ground and is trapped in glaciers can remain there for 20 to 100 years. If the water molecule falls from the atmosphere to the oceans, it will remain there for over 3,000 years. And, if a water molecule descends from the atmosphere and onto an ice sheet on the Antarctic, it could stay in residence for 900,000 years or more.[20]

Along its journey, the water molecule is evaporating from the oceans, lakes, and rivers, as well as from plant leaves via transpiration, and travels upwards to the atmosphere, where it accumulates in the form of clouds. It then falls back onto the land's surface and the rivers, lakes, and oceans of the world, where it takes up residence and eventually transits back again and again in the hydrological cycle, animating the soil, plants, and forests and providing nourishment for all the species on Earth, allowing life to exist in all of its diverse forms.

The workings of Planet Aqua need to be embedded into the heart of our educational system, our cultural lore, our ways of governance, our approach to the economy, our sense of selfhood, and the way we experience life itself. We are of Planet Aqua – it defines us, yet we are largely clueless of this force that animates all of life. What better place to rethink our notions of time, space, and agency than to incorporate water themes into our calendrical life. That means moving beyond the simplistic notion that water is either a holy offering of the gods or, in contemporary times, a passive commercial resource taken up by the capitalist system.

It's likely to surprise most everyone not intimately involved in studying and managing the waters that a "Water Year" formally exists within the scientific community – not as a public relations exercise but as a systemic conceptual tool for tracking the yearly trajectory of the hydrological cycle. The U.S. Geological Survey defines the Water Year in its official reports as a twelve-month period, beginning on October 1 and extending through September 30 of the following year. While the U.S. Geological Survey's Water Year is meant to be a general norm, the exact date of the water year can vary depending on each ecoregion. For example, the Florida Water Year begins on May 1 and runs through April 30. For most regions of the country, October 1 represents the time when the water is generally at its lowest point of the year until the following spring when snow melts and the water flows. Every schoolchild is familiar with the adage, "April showers bring May flowers."

Combining hydrological data via GRACE analytics and algorithms and the Water Year data provides a real-time assessment of the impacts of the hydrological cycle across every region on Earth. Those data, along with data inputs from the World Meteorological Organization and other bodies like the U.S. Geological Survey and their counterparts in other nations should be incorporated into both global and regional hydrological calendars and be up to the moment and readily assessable online, so that the public can be up to date on more than the ten-day weather forecast, but also what's occurring in real time and over time to their regional hydrological cycle and ecosystem, as well as what's happening across the planet.

Back in 1993, the United Nations led the way by inserting a reference about the hydrological cycle into its global calendar, placing "World Water Day" on March 22, so all the nations might celebrate, commemorate, and reaffirm our species' obligation to steward the waters on Planet Aqua. Other hydrological commemorations are quickly finding their way onto regional and global calendars, including "National Ground Water Week," "National Rivers Month," "World Oceans Day," "Lake Appreciation Month," and "Clean Water Week."

CHAPTER 13

Swallowed by the Metaverse or Buoyed by the Aquaverse

Finding our way back to nature will not be an easy journey. By 2022, the average American was spending 92% of his or her day indoors and over seven hours looking at a screen. And, unless we think Americans are an exception, worldwide the average person spends six hours and fifty-seven minutes each day looking at a screen, be it a smartphone, iPad, desktop computer, or TV. Some countries are even more glued to the screen. South Africa tops the list, spending ten hours and 46 minutes a day on the screen. During the recent COVID-19 pandemic, American researchers found that twelve to thirteen-year-olds had doubled "non-school related screen time to 7.7 hours."[1]

And now with Facebook rebranding itself as Meta and the techno world rushing to create virtual worlds that would allow the entire human race to engage in social life and commerce and even new forms of governance virtually, it appears that our species, whether intentionally or not, is curling up in a virtual techno-environment far removed from the prospect of an intimate reaffiliation with the natural world. There is no precedent for this phenomenon in the history of our species on the planet. The natural world of which we are an intimate part, has become the alien other. Or, to reverse the equation, our species is becoming the alien other, choosing to secede from the Earth behind closed doors to be swept away into multiple virtual worlds interacting with our avatar's second self, and even extending our virtual senses to include touch and smell along with sight and hearing.

When we do venture outside, we are increasingly being fitted with augmented virtual reality devices that can overlay data, anecdotes, and messaging onto the physical environment, making it a stage filled with virtual images and streaming data – unconsciously adapting the whole of the outside world to our own utilitarian desires – more or less treating the natural environment as a backdrop to be enhanced or littered, depending on one's outlook.

An Ecological Vision or Dystopian Nightmare: Two Paths to the Future

The metaverse, which is little more than a computer-generated universe, is touted as the ultimate utopian dream, an alternative world whose endgame is the singularity, an immortal space thought up by tech gurus that resides beyond the flesh and blood of existence – what its architects call "virtual reality." Ray Kurzweil, Google's former director of engineering, argues that humans will transcend "the limitations of our biological bodies and brain and that future machines will be human, even if they are not biological."[2] How soon? He says, "I set the date for the singularity as by 2045."[3]

Why the sudden frenzy to escape into virtual worlds and dreams of the singularity? It might have something to do with the dawning awareness that we find ourselves in the sixth extinction of life on Earth and are in a desperate search for salvation, but this time of a more technical kind. And it's the rewilding of the planetary waters that's cowed our species. The sheer power of the hydrosphere conjures up the terror of the sublime. We are each bearing witness to the bitter winter snows, the torrential spring floods, the excruciating summer droughts, heatwaves, and wildfires, and the devastating fall hurricanes. Our planet's ecosystems are collapsing, our built environment is crumbling, and increasing numbers of our human family are dying. We are told that it's going to get worse, and not for a little while but perhaps for millennia to come.

If the sublime is the fear of witnessing up close the overwhelming power unleashed by the Earth's principal agencies, it's the waters that unleash the fury and bend the other great spheres to its will. But we have a hard time coming to terms with the fact that it is our species that lit the match by digging up the burial grounds of the carboniferous era from 350 million years ago and exhuming the dead bodies of plants and animals that had long since morphed into coal, oil, and natural gas and used these fuels to power the Industrial Age. This exhumation is now taking us to a fast-warming climate and an extinction event accompanied by the rewilding of the hydrosphere. How, then, do we respond to the rewilding of the hydrosphere on our water planet?

Were Immanuel Kant alive today, he would, no doubt, view the rewilding waters as an anathema to be cordoned off and very likely heap praise on the metaverse as the ultimate triumph of "detached reason" over the sublime – a force independent of and superior to a flawed physical world of emotions and transient physicality. His followers, the tech wizards of the metaverse, invite all of us to become immersed in a virtual utopia unburdened by the ephemeral nature of earthly existence. They regard the singularity as the final metamorphosis beyond our wretched physical constraints, inviting our species to cast aside the anguish of mortality. Arthur Schopenhauer, on the other hand, were he to revisit the human family at this late hour, might likely regard this moment as the triumph of a different kind – that is, a coming to terms with the sublime by recognizing that our very physicality is eternally embedded, even as it is ephemerally lived. We are each of the universe.

Our every breath changes forever every other condition in the universe, regardless of how small the contribution. The mathematician and meteorologist Edward Norton Lorenz was among the first scientists to recognize that even the slightest fluctuation – for example, a butterfly flapping its wings several weeks earlier, regardless of how minuscule the perturbation, would affect the exact time of a formation of a tornado and the path it would take later on.

While Lorenz's description of the butterfly effect in 1972 captured the public's imagination, other earlier scholars and mathematicians were already weighing in on the subject. In his essay, *The Vocation of Man* (1800), Johann Gottlieb Fichte remarked that "you could not remove a single grain of sand from its place without thereby changing . . . something throughout all parts of the immeasurable whole." In 1950, Alan Turing, the father of artificial intelligence, whose trailblazing work established the operational guidelines for information theory, computation, and ultimately the work that led to the actualization of virtual reality, took note that "the displacement of a single electron by a billionth of a centimeter at one moment might make the difference between a man being killed by an avalanche a year later or escaping."[4]

Every moment of our lived experience – how we live, what we do, how we act, and even the thoughts we share – like the actions of every other creature, ripples out and affects every other phenomenon in a self-organizing and ever-evolving planet. The totality of our personal lived existence is 'immortalized' into the life of the Earth. This is a different kind of eternal being – living on and affecting all of life that comes after us on an animated planet.

Or, to get a bit more intimate, our genetic being and our individual life experiences are preconditioned by the life imprint of every one of our ancestors, be they far removed in time and space or more directly tied to our particular family tree. To make the point even more personal, with every single human body having upward of 4×10^{27} hydrogen atoms and 2×10^{27} oxygen atoms, it's a sure thing that some of those atoms were at one time or another in the bodies of other human beings and other creatures that preceded us throughout history. Likewise, some of those hydrogen and oxygen atoms that were once in our bodies will likely find their way into other human beings and fellow creatures that will come after us.[5]

From a scientific point of view, our bodies are not relatively closed-off autonomous agents but open dissipative systems. Every human body is wrapped up in a semi-permeable membrane, selectively allowing the passing of chemical elements and minerals through its interior – oxygen, hydrogen, nitrogen, carbon, calcium, phosphorus, potassium, sulfur, sodium, chlorine, and so on – that come from across the biosphere.[6] Our bodies, then, are one among the multitude of mediums that host the earth's elements and minerals.

Both the Abrahamic religions and classical Western philosophy perceived immortality very differently as a non-material phenomenon that exists in an atemporal realm far removed from the physicality of unfolding life. In modern times, Immanuel Kant, following the script laid out by Plato and the other Greek philosophers, argued that pure thought and reason are unencumbered by the feelings, emotions, sensibilities, and lived experiences of daily life.

Today's architects of the metaverse – a virtual realm increasingly removed from the raw physicality of lived existence – are merely playing out the most recent reincarnation of longstanding religious and philosophical traditions wedded to the belief in a nonmaterial existence independent of anything that could be characterized as the natural world. This is all to the end of evading death by bypassing the phenomenology of living in real time and in physical space, and experiencing the vulnerabilities and frailties, joys and suffering, awe and wonderment, empathy and compassion, imagination and transcendence, and the many serendipities that accompany being alive.

Alfred North Whitehead parted company with the philosophers and mathematicians of the past, arguing that existence is not to be found in either the heavens above or in the detached rational recesses of the mind but rather in the process of lived experience. Both heaven and pure reason give no credence to lived experience and exist purely in an atemporal realm. Think about it – why would there be any passage of time in heaven which is everlasting life? Recall that Newton's three laws of gravity are absent time's arrow. To be immortal and godlike – the dream of every disciple of the singularity – would be to eclipse time and space all together. To think of God as a spatial/temporal being would be ridiculous.

What's overlooked in all of the discussions of the singularity and transhumanism is what to do about the empathic impulse. How does overcoming time, space, and the phenomenology of lived experience fit in with the empathic drive built into the neurocircuitry of our species, which neuroscientists are beginning to suspect is one of our defining attributes, making us among the most adaptive of species? To transcend corporeality – that is, lived experience in time and space – is to leave empathy behind. Neurobiologists are coming to understand the complex physiology of the empathic drive and the critical role it plays in the sociality of our species and our extraordinary ability to adapt to an ever-evolving and self-actualizing planet.

In 1837, the Royal Danish Society celebrated the life of Immanuel Kant by offering a prize for the best essay on the question:

> Are the source and foundation of morals to be looked for in an idea of morality lying immediately in consciousness (or conscience) and in the analysis of the other fundamental moral concepts springing from that idea, or are they to be looked for in a different ground of knowledge?[7]

Schopenhauer sent in his submission in 1839. His was the only entry. Nonetheless, he was denied the prize. The Royal Danish Society said he had failed to understand the question. But that was a ruse. Their real reason for denying him the prize became clear later on in their explanation. Schopenhauer had dared to suggest, against all the conventional wisdom of the time, that compassion, not pure reason, was the basis of morality and that emotions and feelings animated the compassionate instinct. Sheer heresy. In a last but stinging rebuke, the jurors expressed their displeasure at the abusive manner in which Schopenhauer treated "several distinguished philosophers of recent times."[8] Although they didn't mention specific names, they had Immanuel Kant in mind. Schopenhauer savaged Kant, belittling his purely rational-based, prescriptive ethics as intellectual fantasy, so out of touch with the way moral behavior unfolds in the real world. Like Hume, Schopenhauer believed that reason is the slave of the passions.

Schopenhauer found Kant's idea that moral laws exist *a priori* and are knowable "independently of all inner and outer experience 'resting simply on concepts of pure reason'" without any empirical basis.[9] He noted that Kant rejected the very idea that morality might be an elemental aspect of consciousness and connected to natural feelings "peculiar to human nature," which would give morality an empirical grounding. Kant is very clear on this point. In *The Foundation of the Metaphysics of Morals*, he writes that moral law:

> must not be sought in man's nature (the subjective) or in the circumstances of the world (the objective) . . . here nothing whatever can be borrowed from knowledge relating to man, i.e., from anthropology . . . indeed we must not take it into our heads to try to derive the reality of our moral principle from the particular constitution of human nature.[10]

What we are left with, argued Schopenhauer, is an ethics that exists *a priori* and outside human experience and which is "entirely abstract, wholly insubstantial, and likewise floating about entirely in air."[11]

So, if morality is not an inherent feature of human nature but exists *a priori* and is independent of human nature, what compels someone to be moral? Kant suggests that one acts in a morally responsible way because of "[t]he feeling that it is incumbent on man to obey the moral law . . . from a sense of duty, not from voluntary inclination." Kant dismisses outright any notion of feelings as a basis for morality:

> Feelings of compassion and of tenderhearted sympathy would even be a nuisance to those thinking on the right lines, because they would throw into confusion their well-considered maxims and provoke the desire to be released from these, and to be subject only to legislative reason.[12]

The question then becomes whether there is some other source within the human animal itself that might be the basis of morality.

Schopenhauer counters with a detailed description of moral behavior that he argues is embedded in the very sinew of human nature – with the qualification that it needs to be brought out and nurtured by society if it is to be fully realized. He argues that "compassion" is a fundamental aspect of our human nature. He explains the phenomenon in the following way. In feeling compassion for another:

> I suffer directly with him, I feel his woe just as I ordinarily feel only my own; and, likewise, I directly desire his weal in the same way I otherwise desire my own . . . At every moment we remain clearly conscious that he is the sufferer, not we; and it is precisely in his person, not in ours, that we feel the suffering, to our sorrow. We suffer with him and hence in him; we feel his pain as his, and do not imagine that it is ours.[13]

In this single statement, Schopenhauer becomes the first person in history to succinctly define the empathic process. All that is missing is the term itself. But he goes further, describing not only the mental acrobatics involved in an empathic extension but also the moral frame and action that naturally flow from it. The compassionate predisposition, when fixed on someone's immediate plight, says Schopenhauer, leads to "the immediate participation, independent of all ulterior considerations, primarily in the suffering of another, and thus in the prevention or elimination of it; for all satisfaction and all well-being and happiness consist in this."[14]

On a deeper level, the empathic embrace – the experience of

another's suffering or even their joy as if we were experiencing it in the marrow of our being – are those special moments when we transcend ourselves beyond time and space and exist in a wholly other realm. It's during these interludes that we shed our corporeality and feel the awe of existence. Empathic moments eliminate the *other*. We feel as one for a brief moment in time, fragile and vulnerable, and reaching out to bolster one another's quest to live, thrive, flourish, and to be. In these moments, there are no subjects and objects but only an entanglement of enjoined lives, experiencing each other as one and by extension becoming one with the whole of existence. That these transcendent moments of empathic embrace are the most vivid experiences in our lives is unquestionable. When we look back at our own lives in our later years, we're likely to recall such interludes as the most precious, abiding, and intensely felt experiences. This is Schopenhauer's sublime.

What then of these other worlds we have created – heavens, utopias, and the singularity. None of these imagined existences have room for the empathic impulse. Why would there be any need for empathy in these fictional realms, where vulnerability, suffering, and bursts of joy are absent. In these fictional worlds, the fragility and finitude of existence have been eliminated, and so too moments of sorrow and exuberance and expressions of consolation and acts of compassion – all the conditions that go in hand with being alive. These experiences would be meaningless in heavens, utopias, or in the singularity. So, too, why would there be any feeling of awe, or sense of wonder, or peaking of imagination, or transcendence in these perfect worlds that are timeless and without blemish?

As for artificial intelligence, even the strongest advocates are without an explanation of how empathy fits in a world of zeroes and ones, ubiquitous pixels and the endless stream of data computation, algorithmic agency, and robotic feedback. If anything, these alternative worlds flatten awe, squelch wonderment, stifle the imagination, eliminate transcendence, squash the sublime, and rob our species of the majesty of breathing in existence.

How dim then is Kant's accounting of the sublime? The idea that in the shadow of the sublime and caught up in the terror but also awe of existence, we ought to call on a detached reason removed from the phenomenology of lived experience to subdue, tame, and bend the planetary forces to human agency seems utterly naive.

If that weren't enough, as the saying goes, "there is no such thing as a free lunch." Artificial intelligence comes with its own entropy bill, and it is likely to be a whopper. Like the intensive carbon footprint of mining for Bitcoin, it has recently come to light that AI has an endemic water footprint that is going to have a significant impact on the depletion of global water availability. Any thoughts about AI taking our species to the Singularity are already being tempered by red flags being raised among climate scientists over the water footprint that's likely to slam the brakes on how fast and how far AI computing can take us on a journey to utopia.

A 2023 study conducted by scientists at the University of California at Riverside is jaw-dropping. It's common knowledge that AI servers in giant data centers are "energy hogs," using vast amounts of energy to power their computers, already accounting for upwards of two percent of global electricity usage, and the industry is only just now emerging. It is the water footprint, however, that until recently has all but been ignored. No longer. For example, Google's data centers "withdrew 25 billion liters [of water] and consumed nearly 20 billion liters of scope-1 water for onsite cooling in 2022, the majority of which was potable water." That same year, Microsoft's total water usage – both withdrawal and consumption – increased by 34% compared to 2021.[15]

The study projects that by 2027 the total operational water withdrawal of global AI will top 4.2 to 6.6 billion cubic meters, which is half of the total water withdrawal of the United Kingdom, a nation experiencing increasing drought year-in and year-out (and the U.K. is the sixth largest economy in the world). These figures don't even begin to include the water withdrawal used for manufacturing computer chips.[16]

The manufacture of a single chip requires nearly eight gallons of water. Here's the dilemma. In 2021, 1.15 trillion chips were shipped worldwide and we are only in the early takeoff stages of the AI revolution, which begs the question no one wants to talk about. How much of the fast diminishing pool of freshwater do we give over to growing the metaverse at the expense of further diminishing the already very scarce pool of available water used by our species and fellow creatures. Is anyone listening?[17]

To draw down these projections to a personal level, the study found that ChatGPT-3 consumes a 500ml bottle of water for every ten to fifty responses in a chat conversation. Imagine hundreds of millions and, eventually, much of the human race, plugging into their favorite

Chatbot 24/7, just as they do today on the internet – especially in a world where the amount of water available per person now is one-half of what it was just fifty years ago.[18]

That doesn't mean that AI is a lost cause, but it's most likely to be used more commonly and sparingly in the deployment and management of the emerging Age of Resilience infrastructure, monitoring and managing the communications internet, renewable energy internet, mobility and logistics internet, water internet, and the IoT sensor networks that allow our species to adapt to the increasingly unpredictable swings of the planet's hydrosphere, lithosphere, atmosphere, and biosphere.

So, as to the metaverse, the newest incarnation of utopian bliss . . . how sad. The very prospect of taking the human race out of the natural world and into a virtual facsimile of reality is incomprehensible. In the end, the recoiling from the lifebeat of existence is doomed to fail. Our very physical being is intimately entwined to all the other agencies of the planet – the hydrosphere, lithosphere, atmosphere, and biosphere. As mentioned earlier, our myriad internal biological clocks in every cell, tissue, and organ are temporally entrained to the circadian, tidal, seasonal, and circannual rhythms of the Earth. As well, immersive electromagnetic fields pulse through the cells, organs, and tissues of our bodies and assist in establishing the processes and patterns by which our DNA lines up and is expressed.

That's not to suggest that we abandon virtual worlds altogether but that we do not make them the whole of our world. There will be plenty of need for virtual worlds as playgrounds and experimental stations, and at times safe harbors to weather the storms, droughts, heatwaves, wildfires, and hurricanes to come on a rewilding planet. But, if the metaverse becomes our surrogate world, we will have locked ourselves away in a self-imposed prison of our own making, finding ourselves alone and ultimately exiled and without the wherewithal to survive, no less thrive, in a wildly morphing planet.

For sure, the climate will continue to warm and the Earth's agencies will continue to morph in wholly new ways driven by the hydrological cycle, taking us into a new epoch. But to abandon nature and hide away is to abandon ourselves. We are of this planet and an extension of the hydrosphere, lithosphere, atmosphere, and biosphere and the biomes and ecosystems and the temporal and spatial rhythms that animate life on the blue marble.

In recent decades, biologists and the medical community have belatedly and sparingly turned their attention to how the lack of physical exposure to the natural environment to which every bodily function is entrained has compromised physical and mental health. Even then, much of the research and the implications are largely ignored, if not dismissed altogether. Living in urban enclosures, we are often unaware of how our moods, behavior, and bodily functions are unconsciously affected by our physiological relationship to the environment, especially our mental and physical well-being. Consider, for example, a walk through a forest versus a walk in an urban setting. The forest walk lowers, on average, the salivary cortisol, which is a measurement of stress, by 13.4% upon viewing the forest and 15.8% after the walk, and pulse rate by six percent upon viewing to 3.9% after walking, while also lowering systolic blood pressure. The parasympathetic nervous activity – the feeling of relaxation – increases by 102% after walking, while the feeling of stress decreases by 19.4% after walking. All of these changes occurring in our own body are the result of taking a walk in the woods.[19]

Regarding the broader issue of exposure to nature and degrees of happiness, a mammoth study published in the *Annual Review of Environment and Resources* found that across the board greater immersion in nature correlated with: reduced stress; improved physical and mental health; the lengthening of attention span; improvement in learning; greater imagination; and a sense of greater connectedness.[20] The famed biologist E. O. Wilson of Harvard University was the first to popularize the term biophilia to describe the inherent biological affiliation our species has for other species in our extended evolutionary family and he traced the connection to the inherited biological trait of empathy for all that's alive in our neurocircuitry, as well as attachment behavior between parental figures and offspring.[21]

Unfortunately, young children today, especially in urban and suburban communities, walled off from interaction with our fellow creatures, and the natural environment, are often taught by their parents to be wary and even afraid of other living creatures and to keep their distance, learning from an early age that nature is an alien other. Richard Louv, the author of *Last Child in the Woods*, reports on a conversation with a young boy on why the child preferred playing indoors rather than outdoors – the boy responded "that's where all the electrical outlets are."[22]

Let's be very clear about what role virtual worlds ought to play on a life-affirming planet. Virtual worlds can help us better understand how to adapt to a rewilding hydrological cycle, but the distinction is whether we will continue to force the planet to adapt to our species or whether we will relearn how to adapt our species to nature's requisites with a degree of empathic sophistication, mindfulness, and critical thinking worthy of our continued existence among our fellow creatures.

With this caveat in mind, there is another side to virtual worlds that receives little hype but ultimately will be important in securing our species' future. That is, we will need to employ a smart digital infrastructure – some of which will be virtual – to collect real-time data, and use analytics, GPS, and IoT surveillance to monitor, model, and adapt to the rewilding of the Earth's hydrosphere, lithosphere, atmosphere, and biosphere, and to assess their impacts on the myriad creatures that inhabit the Earth. These smart digital technologies and accompanying infrastructures will be instrumental in employing a new approach to scientific inquiry: one that replaces the worn conventional scientific method of inductive, deductive, and detached reasoning designed to capture and adapt nature to our species and, instead, replaces it with abductive reasoning, entwined with the new complex adaptive social and ecological systems modeling designed to adapt our species to nature's calling. If we do this, we might have a chance to survive a rewilding Earth and acclimate in new ways.

If it's about securing our well-being during the tumultuous change in the Earth's climate, we will have to recognize that there will be extended periods when we will need to be inside conducting our daily life, at least in part, in virtual worlds. That's a given. But it's also true that we will need to brave the new normal on the outside with a steely determination as well as a biophilic sense of how to engage the natural environment.

Where, then, do we go from here? Do we slip into the Kantian rabbit hole of detached reason to tame the Earth or choose the path that Schopenhauer, Goethe, and others have embraced – an empathic and compassionate biophilia attachment with all that is alive so that life might be replenished? Among our species' many attributes is that we are genetically predisposed to a nomadic way of life and have lived so for 95% of the time that we have been on Earth, all the while enduring extreme weather regimes. Learning to live in part in virtual worlds, while re-embedding ourselves back into the planetary spheres, biomes, and ecosystems, and particularly the waters, will determine if and under what conditions our species will survive and flourish on the blue planet. We live on Planet Aqua – that's the simple reality. The hydrological cycle is the prime mover that conditions all the other agencies of the planet. It's the waters from which all of life emerged and to which every living creature is beholden that is the implacable reality of existence on our planet. That acknowledgment is the saving grace by which we will hopefully flourish into the far-distant future.

Planet Aqua: Rebranding Our Home

Renaming our planet Aqua is not just a rhetorical exercise but a change in our species' sense of planetary orientation. Coming to an understanding of the waters as the animating force of life on the planet raises the question of its legal status. If we begin to think of the water less as a "commercial resource" and more as the "life source" on the planet its legal status would need to be completely rethought. That's already beginning to happen. A grassroots movement is taking off across the world to recognize rivers, lakes, seas, and even oceans, as legal persons under the law, with the right to exist, unhampered by human intervention.

If this sounds a bit over the top, it should be remembered that corporations were granted a quasi-legal status as a person in the United States as far back as the nineteenth century under the 14th amendment to the Constitution. Nor is the U.S. alone. The United Kingdom and Canada have also granted corporations many of the same rights as persons.

Ecuador became the first country to incorporate the rights of nature into its constitution in 2008. Mexico, Colombia, and Bolivia have made similar legal changes. Bangladesh, Australia, and New Zealand have gone even further in regard to the legal rights of rivers. In 2017, New Zealand granted the Whanganui River rights as an independent entity, safeguarding its right to exist as "an indivisible whole from source to sea," signaling an extraordinary departure from the way rivers have been seized, rerouted, and blocked from following their natural course. Guardians were even appointed to act on behalf of the river and enforce its right to flow freely in the court of law.[23]

In India, a state high court recently attempted to give the Ganges and Yamuna Rivers legal personhood in 2017. The Supreme Court ruled soon after that rivers cannot be viewed as legal entities. In 2021, local authorities in Quebec, Canada, granted legal personhood to the Magpie River, including the "right to flow, the right to be safe from pollution – and the right to sue" in court. Australia has taken the question of the status of the waters a step further, recognizing the Yarra River as "a living integrated entity," all of which makes sense since there is a growing consensus within the scientific community that the waters are the primordial life force that generates and sustains all living creatures on the planet.[24]

The extension of legal rights to the waters upends the very basis of modern civilization and the capitalist system which views the usurpation, commodification, and consumption of all of nature as the most fundamental of all rights – a legacy handed down over the ages from the Abrahamic God who bequeathed to Adam and his heirs dominion over all the creatures that inhabit the Earth and the whole of the natural world.

Recognizing the waters as the generator of all life and therefore deserving of legal recognition embedded in the constitutions, laws, codes, regulations, and daily activity of every governing jurisdiction and culture marks the most consequential change in our species' relationship to the Earth since the beginning of civilization. In the aftermath of the last Ice Age and the onset of a mild temperate weather regime, we found myriad ways to force the waters to adapt nearly exclusively to our human needs, beginning with the rise of the great urban hydraulic civilizations 6,000 years ago and culminating in the water–energy–food nexus of the fossil fuel-based Industrial Era and the Age of Progress. Now, the waters are breaking away, touched off by the change in the climate of the planet and destroying the very fabric of civilization, forcing our species to change course and make a great turnaround by relearning how to adapt our species to Planet Aqua. Remaking society means re-consecrating our intimate bond to the waters of life. Whether on our planet and likely throughout the universe, water is the generator of all of life.

However, time is running short. Our planet is on life support, and whether we survive or perish into the fossil record will depend on our recognition of and identification with the waters of life. Like it or not, the planet's hydrosphere is searching for a new normal, and every species is caught up in the tumult. If there is a single change that can renew our chances to survive in the epoch to come it is marshaling the collective will to rename our home Planet Aqua and succor the hydrosphere's search for a new cycle. The constant awareness that each of us and our fellow creatures are of the waters will steady us in the long journey ahead. Recognizing that we live and thrive on Planet Aqua will change our species' meta narrative across every realm and prepare us to adapt to a second coming of life on the Blue Marble. Every governing jurisdiction should be pressed to embed a second official naming of our sphere as Planet Aqua in their constitution, covenants, and bylaws. Our educational system will also need to incorporate the idea of living on Planet Aqua in pedagogy and across the curricula, from the vantage point of immersion in all that constitutes the sublime waters of an animated planet.

We are closing in on an endgame for our species and the continuance of life itself on Earth. We are at worst asleep and at best in a deep fog, hoping for a miracle – some kind of unanticipated dramatic and miraculous turnaround in the climate that was never factored into the reams of studies and reports by our climate scientists. This pollyannaish hope is a recipe for inaction. Nor can we any longer hide behind the meek and hastily cobbled-together climate initiatives that have proved to be too little to address the enormity of the planetary crisis before us.

Today, and moving into the far-distant future, climate change will govern the rules of the game. Every moment will conjure up the sublime in all sorts of ways to which our species and our fellow creatures will have to continuously adapt. And at every juncture before us, it will be the hydrosphere that conditions our choices. Whether we embrace the sublime and readapt to the waters with a sense of awe, wonder, imagination, and empathic attachment or huddle in fear and lockdown will decide our fate and likely the fate of our fellow creatures that make up our evolutionary family. Resist and cower or ride the waves wherever they take Planet Aqua.

Notes

Introduction

1 "Biologists Think 50% of Species Will Be Facing Extinction by the End of the Century," *Guardian*, February 25, 2017. https://www.theguardian.com/environment/2017/feb/25/half-all-species-extinct-end-century-vatican-conference.

2 Kevin E. Trenberth, "Changes in Precipitation with Climate Change," *Climate Research* 47 (1), March 31, 2011: 123–38. https://doi.org/10.3354/cr00953.

3 "Ecological Threat Register 2020: Understanding Ecological Threats, Resilience and Peace," Sydney: The Institute for Economics & Peace, September 2020: 38. https://reliefweb.int/report/world/ecological-threat-register-2020-understanding-ecological-threats-resilience-and-peace.

4 Ibid., 2.

5 Ibid., 4.

6 Ibid., 2.

7 Ibid., 3–52.

8 "State of the Climate: Monthly Drought Report for May 2022," National Centers for Environmental Information, June 2022. https://www.ncei.noaa.gov/access/monitoring/monthly-report/drought/202205.

9 "International Drought Resilience Alliance: UNCCD." IDRA. https://idralliance.global/. Sengupta, Somini. "Drought Touches a Quarter of Humanity, U.N. Says, Disrupting Lives Globally." *The New York Times*, January 11, 2024. https://www.nytimes.com/2024/01/11/climate/global-drought-food-hunger.html.

10 Jeff Masters, "Death Valley, California, Breaks the All-Time World Heat Record for the Second Year in a Row," Yale Climate Connections, July 12, 2021. https://yaleclimateconnections.org/2021/07/death-valley-california-breaks-the-all-time-world-heat-record-for-the-second-year-in-a-row.

11 "Eight Climate Change Records the World Smashed in 2021," World Economic Forum, May 18, 2022. https://www.weforum.org/agenda/2022/05/8-climate-change-records-world-2021/.

12 Margaret Osborne, "Earth Faces Hottest Day Ever Recorded – Three Days in a Row," Smithsonian.com, July 6, 2023. https://www.smithsonianmag.com/smart-news/earth-faces-hottest-day-ever-recorded-three-days-in-a-row-180982493/.

13 "National Fire News," National Interagency Fire Center, September 2023. https://www.nifc.gov/fire-information/nfn.
14 Rebecca Falconer, "Canada's Historic Wildfire Season Abates after 45.7 Million Acres Razed." *Axios*, October 20, 2023. https://www.axios.com/2023/10/20/canada-record-2023-wildfire-season-end.
15 "Wildfire Graphs," CIFFC Canadian Interagency Forest Fire Center. https://ciffc.net/statistics; "Giant Carbon Shield," Boreal Conservation. https://www.borealconservation.org/giant-carbon-shield.
16 "Ecological Threat Register 2020," op. cit. p. 13.
17 "Ecological Threat Report 2021: Understanding Ecological Threats, Resilience, and Peace," Sydney: The Institute for Economics & Peace, October 2021: 4. https://www.economicsandpeace.org/wp-content/uploads/2021/10/ETR-2021-web.pdf.
18 Aylin Woodward and Marianne Guenot,"The Earth Has Tilted on its Axis Differently over the Last Few Decades Due to Melting Ice Caps." *Business Insider*, March 21, 2023. https://www.businessinsider.com/earth-axis-shifted-melting-ice-climate-change-2021-4.
19 Laura Poppick, "The Ocean Is Running out of Breath, Scientists Warn," *Scientific American,* March 20, 2019. https://www.scientificamerican.com/article/the-ocean-is-running-out-of-breath-scientists-warn/.
20 "New Study: U.S. Hydropower Threatened by Increasing Droughts Due to Climate Change," WWF, February 24, 2022. https://www.worldwildlife.org/press-releases/new-study-us-hydropower-threatened-by-increasing-droughts-due-to-climate-change#:~:text=The%20study%20finds%20that%20by,from%201%20in%2025%20today.
21 "Learn about Our Great Lakes," SOM – State of Michigan. https://www.michigan.gov/egle/public/learn/great-lakes#:~:text=The%20combined%20lakes%20contain%20the,economy%2C%20society%2C%20and%20environment.
22 Antonio Zapata-Sierra, Mila Cascajares, Alfredo Alcayde, and Francisco Manzano-Agugliaro, "Worldwide Research Trends on Desalination," *Desalination*. 2021. https://doi.org/10.1016/j.desal.2021.115305.
23 Stephanie Rost, "Navigating the Ancient Tigris – Insights into Water Management in an Early State," *Journal of Anthropological Archaeology,* 54, 2019: 31–47, SSN 0278–4165; also "Water Management in Mesopotamia from the Sixth till the First Millennium BCE," WIREs Water e1230, 2017. doi:10.1002/wat2.1230; S. Mantellini, V. Picotti, A. Al-Hussainy, N. Marchett, F. Zaina, "Development of Water Management Strategies in Southern Mesopotamia during the Fourth and Third Millennium BCE," *Geoarchaeology*, 2024: 1–32. https://doi.org/10.1002/gea.21992.
24 Karl W. Butzer, "Early Hydraulic Civilization in Egypt: A Study in Cultural Ecology." University of Chicago, 1976. https://isac.uchicago.edu/sites/default/files/uploads/shared/docs/early_hydraulic.pdf.
25 Pushpendra Kumar Singh, Pankaj Dey, Sharad Kumar Jain, and Pradeep P. Mujumdar, "Hydrology and Water Resources Management in Ancient India," *Hydrology and Earth System Sciences* 24, 2020: 4691–4707. https://doi.org/10.5194/hess-24-4691-2020.
26 Bin Liu et al., "Earliest Hydraulic Enterprise in China, 5100 Years Ago," *Proceedings of the National Academy of Sciences of the United States of America*, 114 (52), 2017: 13637–13642. https://doi.org/10.1073/pnas.1710516114.

27 Andrew Wilson, "Water, Power and Culture in the Roman and Byzantine Worlds: An Introduction," *Water History* 4 (1), March 28, 2012: 1–9. https://doi.org/10.1007/s12685-012-0050-2; Christer Bruun, "Roman Emperors and Legislation on Public Water Use in the Roman Empire: Clarifications and Problems," *Water History* 4 (1), March 28, 2012: 11–33. https://doi.org/10.1007/s12685-012-0051-1; Edmund Thomas, "Water and the Display of Power in Augustan Rome: The So-Called 'Villa Claudia' at Anguillara Sabazia," *Water History* 4 (1), March 28, 2012: 57–78. https://doi.org/10.1007/s12685-012-0055-x.
28 Alessandro F. Rotta Loria, "The Silent Impact of Underground Climate Change on Civil Infrastructure," *Communications Engineering* 2 (44), 2023. https://doi.org/10.1038/s44172-023-00092-1.
29 Adam Zewe, "From Seawater to Drinking Water, with the Push of a Button," *MIT News*, Massachusetts Institute of Technology. https://news.mit.edu/2022/portable-desalination-drinking-water-0428.

1 First There Was the Waters

1 Harvey, Warren Zev. "Creation from Primordial Matter: Did Rashi Read Plato's Timaeus?" Thetorah.com, 2019. https://www.thetorah.com/article/creation-from-primordial-matter-did-rashi-read-platos-timaeus.
2 Ibid.
3 Damien Carrington, "Climate Crisis Has Shifted the Earth's Axis, Study Shows," *Guardian*, Guardian News and Media, April 23, 2021. https://www.theguardian.com/environment/2021/apr/23/climate-crisis-has-shifted-the-earths-axis-study-shows.
4 Aylin Woodward, "The Earth Has Tilted on its Axis Differently over the Last Few Decades Due to Melting Ice Caps." March 23, 2023. https://africa.businessinsider.com/science/the-earth-has-tilted-on-its-axis-differently-over-the-last-few-decades-due-to-melting/jntq86j.
5 Stephanie Pappas, "Climate Change Has Been Altering Earth's Axis for at Least 30 Years," *LiveScience*. Future U.S. Inc, April 28, 2021. https://www.livescience.com/climate-change-shifts-poles.html.
6 Hoai-Tran Bui, "Water Discovered Deep beneath Earth's Surface," *USA Today*, June 12, 2014. https://www.usatoday.com/story/news/nation/2014/06/12/water-earth-reservoir-science-geology-magma-mantle/10368943/.
7 Juan Siliezar, "Harvard Scientists Determine Early Earth May Have Been a Water World." Harvard Gazette, November 9, 2023. https://news.harvard.edu/gazette/story/2021/04/harvard-scientists-determine-early-earth-may-have-been-a-water-world/.
8 Water Science School, "The Water in You: Water and the Human Body," U.S. Geological Survey, May 22, 2019, https://www.usgs.gov/special-topic/water-science-school/science/water-you-water-and-human-body?qt-science_center_objects=0#qt-science_center_objects.
9 H. H. Mitchell, T. S. Hamilton, F. R. Steggerda, and H. W. Bean, "The Chemical Composition of the Adult Human Body and Its Bearing on the Biochemistry of Growth," *Journal of Biological Chemistry* 158 (3), May 1, 1945: 625–637. https://www.jbc.org/article/S0021-9258(19)51339-4/pdf.

10 "What Does Blood Do?" Institute for Quality and Efficiency in Health Care, InformedHealth.org, U.S. National Library of Medicine, August 29, 2019. https://www.ncbi.nlm.nih.gov/books/NBK279392/.
11 Ibid., "The Water in You," *Water Science School*. See note 8.
12 "Biologists Think 50% of Species Will Be Facing Extinction by the End of the Century," *Guardian*, February 25, 2017. https://www.theguardian.com/environment/2017/feb/25/half-all-species-extinct-end-century-vatican-conference.
13 Vivek V. Venkataraman, Thomas S. Kraft, Nathaniel J. Dominy, Kirk M. Endicott, "Hunter Gatherer Residential Mobility and the Marginal Value of Rainforest Patches," *Proceedings of the National Academy of Sciences* 114 (12), March 6, 2017: 3097. https://www.pnas.org/doi/10.1073/pnas.1617542114.
14 Kat So and Sally Hardin, "Extreme Weather Cost U.S. Taxpayers $99 Billion Last Year, and It Is Getting Worse," Center for American Progress, September 1, 2021. https://www.americanprogress.org/article/extreme-weather-cost-u-s-taxpayers-99-billion-last-year-getting-worse/; National Oceanic and Atmospheric Administration, "Billion-Dollar Weather and Climate Disasters: Events." https://www.ncdc.noaa.gov/billions/events/US/2020.
15 Adam B. Smith, "2021 U.S. Billion-Dollar Weather and Climate Disasters in Historical Context." January 23, 2022 https://www.climate.gov/news-features/blogs/beyond-data/2021-us-billion-dollar-weather-and-climate-disasters-historical.
16 Sarah Kaplan and Andrew Ba Tran, "Over 40% of Americans Live in Counties Hit by Climate Disasters in 2021," *Washington Post*, January 5, 2022. https://www.washingtonpost.com/climate-environment/2022/01/05/climate-disasters-2021-fires/; Alicia Adamczyk, "Here's What's in the Democrats' $1.75 Trillion Build Back Better Plan," CNBC, October 28, 2021. https://www.cnbc.com/2021/10/28/whats-in-the-democrats-1point85-trillion-dollar-build-back-better-plan.html.
17 "Dams Sector," Cybersecurity and Infrastructure Security Agency. https://www.cisa.gov/topics/critical-infrastructure-security-and-resilience/critical-infrastructure-sectors/dams-sector; "Dams," ASCE's 2021 Infrastructure Report Card, July 12, 2022. https://infrastructurereportcard.org/cat-item/dams-infrastructure/; "Levees." ASCE's 2021 Infrastructure Report Card, July 12, 2022. https://infrastructurereportcard.org/cat-item/levees-infrastructure/.
18 Jeffrey J. Opperman, Rafael R. Camargo, Ariane Laporte-Bisquit, Christiane Zarfl, and Alexis J. Morgan, "Using the WWF Water Risk Filter to Screen Existing and Projected Hydropower Projects for Climate and Biodiversity Risks," *Water* 14 (5), 2022: 721. https://doi.org/10.3390/w14050721.
19 Mateo Jasmine Munoz, "Lawrence Joseph Henderson: Bridging Laboratory and Social Life." DASH Home, 2014. https://dash.harvard.edu/handle/1/12274511.
20 "Gaia hypothesis." *Encyclopedia Britannica*, April 20, 2023. https://www.britannica.com/science/Gaia-hypothesis.
21 Madeleine Nash, "Our Cousin the Fishapod," *TIME Magazine*, April 10, 2006. http://content.time.com/time/magazine/article/0,9171,1181611,00.html.
22 Ibid.
23 Xupeng Bi, Kun Wang, Liandong Yang, Hailin Pan, Haifeng Jiang, Qiwei Wei, Miaoquan Fang et al., "Tracing the Genetic Footprints of Vertebrate Landing

in Non-Teleost Ray-Finned Fishes," *Cell* 184 (5), February 4, 2021: 1377–91. https://doi.org/10.1016/j.cell.2021.01.046.

24 Ibid.

25 X. Y. Sha, Z. F. Xiong, H. S. Liu, X. D. Di, and T. H. Ma, "Maternal–Fetal Fluid Balance and Aquaporins: From Molecule to Physiology." *Acta Pharmacologica Sinica* 32(6), 2011: 716–720. https://doi.org/10.1038/aps.2011.59.

26 From the Epic of Gilgamesh. https://www.cs.utexas.edu/~vl/notes/gilgamesh.html.

27 Mircea Eliade, *Patterns in Comparative Religion*. Lincoln: University of Nebraska Press, 1996, pp. 188–189.

28 Ivan Illich, *H_2O and the Waters of Forgetfulness*. New York: Marion Boyars, 1986, pp. 24–25.

29 United Nations, "Fact Sheet," The Ocean Conference, June 5–9, 2017, https://sustainabledevelopment.un.org/content/documents/Ocean_Factsheet_People.pdf.

30 M. Kummu, H. de Moel, P. J. Ward, and O. Varis, "How Close Do We Live to Water? A Global Analysis of Population Distance to Freshwater Bodies," *PLOS ONE* 6(6), 2011: e20578. https://doi.org/10.1371/journal.pone.0020578.

31 Sebastian Volker and Thomas Kistemann. "The Impact of Blue Space on Human Health and Well-Being – Salutogenetic Health Effects of Inland Surface Waters: A Review," *International Journal of Hygiene and Environmental Health* 214 (6), 2011: 449–460. https://doi.org/10.1016/j.ijheh.2011.05.001.

32 U. Nanda, S. L. Eisen, and V. Baladandayuthapani, "Undertaking an Art Survey to Compare Patient Versus Student Art Preferences," *Environment and Behavior*, 40(2), 2008: 269–301. https://doi.org/10.1177/0013916507311552.

33 Jenna Karjeski, "This is Water," *The New Yorker*, September 19, 2008, https://www.newyorker.com/books/page-turner/this-is-water.

34 Shmuel Burmil, Terry C. Daniel, and John D. Hetherington, "Human Values and Perceptions of Water in Arid Landscapes," *Landscape and Urban Planning* 44 (2–3), 1999: 99–109, ISSN 0169 2046, https://doi.org/10.1016/S0169-2046(99)00007-9.

35 Ibid., 100.

36 Eugene C. Robertson, " The Interior of the Earth," *Geological Survey Circular*, 1966. https://pubs.usgs.gov/gip/interior/.

37 Megan Fellman, "New Evidence for Oceans of Water Deep in the Earth." *Northwestern Now*, 2014. https://news.northwestern.edu/stories/2014/06/new-evidence-for-oceans-of-water-deep-in-the-earth/.

38 Steve Nadis, "The Search for Earth's Underground Oceans," *Discover Magazine*, June 13, 2020. https://www.discovermagazine.com/planet-earth/the-search-for-earths-underground-oceans.

39 Megan Fellman, "New Evidence for Oceans of Water Deep in the Earth." See note 37.

40 Steve Nadis, "The Search for Earth's Underground Oceans." See note 38.

41 Ibid.

42 Ibid.

2 The Earth Be Dammed: The Dawn of Hydraulic Civilization

1 D. Koutsoyiannis and A. Angelakis, "Agricultural Hydraulic Works in Ancient Greece," *Encyclopedia of Water Science*, 2004. https://doi.org/10.1081/E-EWS 120020412; Damian Evans, "Hydraulic Engineering at Angkor," *Encyclopaedia of the History of Science, Technology, and Medicine in Non-Western Cultures*, 2016, pp. 2215–2219. https://doi.org/10.1007/978-94-007-7747-7_9842; Yolanda López-Maldonado, "Little Has Been Done to Recognise Ancient Mayan Practices in Groundwater Management," UNESCO.org, May 3, 2023. https://www.unesco.org/en/articles/little-has-been-done-recognise-ancient-mayan-practices-groundwater-management; UN-Water Summit on Groundwater, Paris; L. F. Mazadiego, O. Puche, and A. M. Hervás, "Water and Inca Cosmogony: Myths, Geology and Engineering in the Peruvian Andes," *Geological Society, London, Special Publications* 310 (1), January 2009: 17–24. https://doi.org/10.1144/sp310.3.

2 Sinclair Hood, *The Minoans: Crete in the Bronze Age*. London, 1971; Mark Cartwright, "Food and Agriculture in Ancient Greece," *World History Encyclopedia*, February 4, 2024. https://www.worldhistory.org/article/113/food-agriculture-in-ancient-greece/; "Ancient Egyptian Agriculture," Food and Agriculture Organization of the United Nations. https://www.fao.org/country-showcase/item-detail/en/c/1287824/#:~:text=The%20Egyptians%20grew%20a%20variety,wheat%2C%20grown%20to%20make%20bread; "Maize: The Epicenter of Maya Culture," Trama Textiles, Women's Weaving Cooperative, February 1, 2019. https://tramatextiles.org/blogs/trama-blog/maize-the-epicenter-of-maya-culture; UNESCO World Heritage. "Angkor," UNESCO World Heritage Centre. https://whc.unesco.org/en/list/668/.

3 Robert K. Logan, *The Alphabet Effect: The Impact of the Phonetic Alphabet on the Development of Western Civilization*, New York: William Morrow, 1986, pp. 60–61.

4 David Diringer, *The Alphabet: A Key to the History of Mankind*, 2nd edn., New York: Philosophical Library, 1953; Ignace Gelb, *A Study of Writing*, revd. edn., Chicago: University of Chicago Press, 1963.

5 Logan, *The Alphabet Effect*, pp. 67–69.

6 L. A. White, *The Evolution of Culture: The Development of Civilization to the Fall of Rome*, Walnut Creek, CA: Left Coast Press, 2007, p. 356.

7 Logan, *The Alphabet Effect*, p. 70.

8 Ibid., 32.

9 Ibid., 78.

10 Jean-Claude Debeir, Jean-Paul Deléage, and Daniel Hémery, *In the Servitude of Power: Energy and Civilization Through the Ages*. John Barzman, trans. London: Zed Books, 1991. p. 21.

11 Lewis Mumford, *The Transformations of Man*, Gloucester, MA: Peter Smith, 1978. p. 40.

12 Karl A. Wittfogel, *Oriental Despotism: A Comparative Study of Total Power*, New York: Vintage Books, 1981, pp. 254–255.

13 Ibid., p. 37.

14 Hannah Ritchie and Max Roser, "Urbanization." Our World in Data. Global Change Data Lab, June 13, 2018. https://ourworldindata.org/urbanization.

15 Rep. *World Population Prospects: The 2014 Revision*, 2015. https://population.un.org/wup/publications/files/wup2014-report.pdf.
16 Joram Mayshar, Omer Moav, and Luigi Pascali. "The Origin of the State: Land Productivity or Appropriability?" *Journal of Political Economy* 130 (4), April 1, 2022. https://doi.org/10.1086/718372.
17 Ibid., 2.
18 Ibid., 27.
19 Ibid., 37.
20 S. Sunitha, A. U. Akash, M. N. Sheela, and J. Suresh Kumar. "The Water Footprint of Root and Tuber Crops." *Environment, Development and Sustainability*, January 26, 2023. https://doi.org/10.1007/s10668-023-02955-1.
21 Cynthia Bannon, *Gardens and Neighbors: Private Water Rights in Roman Italy*. University of Michigan Press, 2009, pp. 65–73; Andrew Wilson, "Water, Power and Culture in the Roman and Byzantine Worlds: An Introduction." *Water History* 4 (1), March 28, 2012: 1–9. https://doi.org/10.1007/s12685-012-0050-2.; Christer Bruun, "Roman Emperors and Legislation on Public Water Use in the Roman Empire: Clarifications and Problems." *Water History* 4 (1), March 28, 2012: 11–33. https://doi.org/10.1007/s12685-012-0051-1.; Edmund Thomas, "Water and the Display of Power in Augustan Rome: The So-Called 'Villa Claudia' at Anguillara Sabazia." *Water History* 4, (1), March 28, 2012: 57–78. https://doi.org/10.1007/s12685-012-0055-x.
22 David Graeber and David Wengrow, *The Dawn of Everything: A New History of Humanity*, New York: Picador/Farrar, Straus and Giroux, 2023.
23 Karl A. Wittfogel, *Oriental Despotism: A Comparative Study of Total Power*, New York: Vintage Books, 1981.
24 Robert L. Carneiro, A *Theory of the Origin of the State*, Institute for Human Studies, 1977.
25 Margaret T. Hodgen, "Domesday Water Mills," *Antiquity* 13 (51), September 1939: 266.
26 Jean-Claude Debeir, Jean-Paul Deléage, and Daniel Hémery, *In the Servitude of Power: Energy and Civilization through the Ages*, London: Zed Books, 1991, p. 75.
27 Ibid., p. 76.
28 Ibid., p. 90.
29 Lynn White, *Medieval Technology and Social Change*, London: Oxford University Press, 1962, pp. 128–129.
30 Piotr Steinkeller, "Labor in the Early States: An Early Mesopotamian Perspective," Institute for the Study of Long-term Economic Trends and the International Scholars Conference on Ancient Near Eastern Economies, 2005. https://www.academia.edu/35603966/Labor_in_the_Early_States_An_Early_Mesopotamian_Perspective.
31 Ibid., 1.
32 Ibid., 2.; "Early History to 17th Century: History of Accounting, A Resource Guide," https://guides.loc.gov/history-of-accounting/practice/early-history.
33 Ibid., 13.
34 F. Krausmann et al., "Global Human Appropriation of Net Primary Production Doubled in the 20th Century," *PNAS* 110 (25), June 2013: 10324–10329.

https://doi.org/10.1073/pnas.1211349110; "What Is Net Primary Productivity for Earth?" Earth How, September 24, 2023; https://earthhow.com/net-primary-productivity/.

35 Lena Hommes, Jaime Hoogesteger, and Rutgerd Boelens, "(Re)Making Hydrosocial Territories: Materializing and Contesting Imaginaries and Subjectivities through Hydraulic Infrastructure," *Political Geography*, ScienceDirect, July 1, 2022, https://www.sciencedirect.com/science/article/pii/S0962629822001123.

36 Ibid., 4.

37 Ibid.

38 Ibid., 5.

39 Ibid., 6.

40 Alice Tianbo Zhang, Johannes Urpelainen, and Wolfram Schlenker. "Power of the River: Introducing the Global Dam Tracker (GDAT)," Center on Global Energy Policy at Columbia University, SIPA, November 2018. https://www.energypolicy.columbia.edu/sites/default/files/pictures/GlobalDams_CGEP_2018.pdf.

41 Ibid., 2.

42 Robyn White, "Lake Mead: Where Does It Get Its Water and Is It Filling up?" *Newsweek*, February 24, 2023. https://www.newsweek.com/lake-mead-water-filling-colorado-explained-reservoir-1783553.

43 William E. Smithe, *The Conquest of Arid America*, Washington, D.C.: Library of Congress, 1900.

44 A. R. Turton, R. Meissner, P. M. Mampane, and O. Seremo, "A Hydropolitical History of South Africa's International River Basins," Report to the Water Research Commission. Pretoria: African Water Issues Research Unit (AWIRU), University of Pretoria, 2004.

45 "In Memoriam A. H. H. OBIIT MDCCCXXXIII," Representative Poetry Online, University of Toronto Libraries, 1908. https://rpo.library.utoronto.ca/content/memoriam-h-h-obiit-mdcccxxxiii-all-133-poems.

46 Christopher Bertram, "Jean-Jacques Rousseau," *Stanford Encyclopedia of Philosophy*, May 26, 2017. https://plato.stanford.edu/entries/rousseau/.

47 Nicolas de Condorcet, "The Progress of the Human Mind." https://wwnorton.com/college/history/ralph/workbook/ralprs24d.htm.

48 Brett Bowden, "Civilization and Its Consequences," *Oxford Academic*, February 11, 2016. https://doi.org/10.1093/oxfordhb/9780199935307.013.30.

49 Ibid., 5; John Stuart Mill, "Essays on Politics and Society," ed. J. M. Robson. The Online Library of Liberty, 1977. https://competitionandappropriation.econ.ucla.edu/wp-content/uploads/sites/95/2016/06/EssaysPolSoc1OnLiberty.pdf.

50 Ibid., 7; François Guizot, *The History of Civilization in Europe*, ed. William Hazlitt, Penguin, 1997.

51 "Franklin Delano Roosevelt, Boulder Dam Dedication Speech, September 30, 1935," *Energy History*, Yale University, September 30, 1935. https://energyhistory.yale.edu/library-item/franklin-delano-roosevelt-boulder-dam-dedication-speech-sept-30-1935.

52 "Annual Freshwater Withdrawals, Agriculture (% of Total Freshwater Withdrawal)," World Bank Open Data. https://data.worldbank.org/indicator/er.h2o.fwag.zs; Dave Berndtson, "As Global Groundwater Disappears, Rice, Wheat and Other International Crops May Start to Vanish," *PBS*, April 17,

2017. https://www.pbs.org/newshour/science/global-groundwater-disappears-rice-wheat-international-crops-may-start-vanish.

53 Ibid.

54 Charles Killinger, *The History of Italy*, Westport, CT: Greenwood Press, 2002, p. 1; Massimo D'Azeglio, *Miei Ricordi* (1891), p. 5.

55 Johann Wolfgang von Goethe, *Werke, Briefe und Gespräche. Gedenkausgabe*, 24 vols. *Naturwissenschaftliche Schriften*, vols. 16–17, ed. Ernst Beutler, Zurich: Artemis, 1948–1953, pp. 921–923.

56 Ibid.

57 Goethe, *Werke, Briefe und Gespräche. Dichtung und Wahrheit*, vol. 10, p. 425.

58 Ibid.

3 Gender Wars: The Struggle Between *Terra Firma* and Planet Aqua

1 Carol P. Christ, "Women Invented Agriculture, Pottery, and Weaving and Created Neolithic Religion," May 18, 2020. https://feminismandreligion.com/2020/05/11/women-invented-agriculture-pottery-and-weaving-by-carol-p-christ/.

2 W. J. MacLennan and W. I. Sellers, "Ageing Through the Ages," *Proceedings of the Royal College of Physicians*, Edinburgh 1999, 29: 71.

3 Veronica Strang, "Lording It over the Goddess: Water, Gender, and Human–Environmental Relations," *Journal of Feminist Studies in Religion* 30 (1), 2014: 85–109.

4 Ibid.

5 Mina Nakatani, "The Myth of Typhon Explained," Grunge, November 9, 2021, https://www.grunge.com/655559/the-myth-of-typhon-explained/.

6 W. Young, and L. DeCosta, "Water Imagery in Dreams and Fantasies," *Dynamic Psychotherapy* 5 (1), Spring/Summer, 1987: 67–76.

7 Veronica Strang, *The Meaning of Water*, Routledge, 2014, p. 86.

8 Veronica Strang, 2005. "Taking the Waters: Cosmology, Gender and Material Culture in the Appropriation of Water Resource," in *Water, Gender and Development*, eds. A. Coles and T. Wallace, Oxford, New York: Berg, 2005, p. 24.

9 Ibid., 31–32.

10 Charles Sprawson, *Haunts of the Black Masseur: The Swimmer as Hero*, London: Vintage Classic, 2018.

11 Ibid., 9.

12 Ibid.

13 Ibid., 15.

14 Ibid.

15 "The Gods and Goddesses of Ancient Rome," *National Geographic*, October 19, 2023. https://education.nationalgeographic.org/resource/gods-and-goddesses-ancient-rome/.

16 Emily Holt, *Water and Power in Past Societies*, Institute for European and Mediterranean Archaeology Distinguished Monograph Series, pp. 118–119. SUNY Press, 2018.

17 Ibid.

18 "The Roman Empire: Why Men Just Can't Stop Thinking About It," *Guardian*, September 19, 2023. https://www.theguardian.com/lifeandstyle/2023/sep/19/the-roman-empire-why-men-just-cant-stop-thinking-about-it.

19 Charles Sprawson, *In Haunts of the Black Masseur: The Swimmer as Hero*. See note 10.
20 Ibid.
21 Claire Colebrook, "Blake and Feminism: Romanticism and the Question of the Other," *Blake/An Illustrated Quarterly*, 2000. https://bq.blakearchive.org/34.1.colebrook.
22 Ibid.
23 Donald Worster, *Nature's Economy*, Cambridge University Press, 1977, p. 30.
24 Claire Colebrook, "Blake and Feminism."
25 Samuel Baker, *Written on the Water: British Romanticism and the Maritime Empire of Culture*. Charlottesville, University of Virginia Press, 2010, p. 14.
26 Ibid.
27 Béatrice Laurent, *Water and Women in the Victorian Imagination*, Oxford: Peter Lang, 2021, pp. 72–73.
28 Ibid.
29 Elaine Showalter, *The Female Malady: Women, Madness, and English Culture, 1830–1980*, London: Virago Press, 1985.
30 "2.1 Billion People Lack Safe Drinking Water at Home, More than Twice as Many Lack Safe Sanitation." World Health Organization, July 12, 2017. https://www.who.int/news/item/12-07-2017-2-1-billion-people-lack-safe-drinking-water-at-home-more-than-twice-as-many-lack-safe-sanitation.
31 Bethany Caruso, "Women Still Carry Most of the World's Water," *The Conversation* U.S., Inc., July 16, 2017. https://theconversation.com/women-still-carry-most-of-the-worlds-water-81054.
32 Jody Ellis, "When Women Got the Right to Vote in 50 Countries," *Stacker*, September 15, 2022. https://stacker.com/world/when-women-got-right-vote-50-countries.
33 Margreet Zwarteveen, "Men, Masculinities and Water Powers in Irrigation," *Water Alternatives* 1 (1), 2008: 114. https://www.water-alternatives.org/index.php/allabs/19-a-1-1-7/file.
34 Judy Wajcman, "Feminism Confronts Technology," Wiley.com, September 2, 1991. https://www.wiley.com/en-us/Feminism+Confronts+Technology-p-9780745607788.
35 Margreet Zwarteveen, "Men, Masculinities and Water Powers in Irrigation."
36 Kuntala Lahiri-Dutt, *Fluid Bonds: Views on Gender and Water*, Stree Books, 2006, p. 44.
37 Ibid., 30.

4 The Paradigmatic Transformation from Capitalism to Hydroism

1 Martin, A. Delgado, "Water for Thermal Power Plants: Understanding a Piece of the Water–Energy Nexus," *Global Water Forum*, June 22, 2015. https://globalwaterforum.org/2015/06/22/water-for-thermal-power-plants-understanding-a-piece-of-the-water-energy-nexus/; A. Delgado, "Water Footprint of Electric Power Generation: Modeling its Use and Analyzing Options for a Water-Scarce Future," Massachusetts Institute of Technology, Cambridge, MA, 2012.
2 "Summary of Estimated Water Use in the United States in 2015," Fact Sheet

2018–3035, U.S. Department of the Interior, June 2018. https://pubs.usgs.gov/fs/2018/3035/fs20183035.pdf.

3 James Kanter, "Climate Change Puts Nuclear Energy into Hot Water," *New York Times*, May 20, 2007. https://www.nytimes.com/2007/05/20/health/20iht-nuke.1.5788480.html; "Cooling Power Plants: Power Plant Water Use for Cooling," World Nuclear Association, September 2020. https://world-nuclear.org/information-library/current-and-future-generation/cooling-power-plants.aspx; "Nuclear Power Plants Generated 68% of France's Electricity in 2021," Homepage – U.S. Energy Information Administration (EIA). https://www.eia.gov/todayinenergy/detail.php?id=55259.

4 Forrest Crellin, "High River Temperatures to Limit French Nuclear Power Production" Reuters, July 12, 2023. https://www.reuters.com/business/energy/high-river-temperatures-limit-french-nuclear-power-production-2023-07-12/.

5 "Water in Agriculture," The World Bank, October 5, 2022. https://www.worldbank.org/en/topic/water-in-agriculture. "Water Scarcity," World Wildlife Fund (WWF). https://www.worldwildlife.org/threats/water-scarcity; "Farms Waste Much of World's Water," *Wired*, Condé Nast, March 19, 2006. https://www.wired.com/2006/03/farms-waste-much-of-worlds-water/.

6 Martin C. Heller and Gregory A. Keoleian. *Life Cycle-Based Sustainability Indicators for Assessment of the U.S. Food System*. Ann Arbor, MI: Center for Sustainable Systems, University of Michigan, 2000, p. 42.

7 Alena Lohrmann, Javier Farfan, Upeksha Caldera, Christoph Lohrmann, and Christian Breyer, "Global Scenarios for Significant Water Use Reduction in Thermal Power Plants Based on Cooling Water Demand Estimation Using Satellite Imagery." LUT University, 2019. DOI: 10.1038/s41560-019-0501-4

8 "All Renewable Power Could Mean 95 Percent Cut in Water Consumption," Water Footprint Calculator, September 9, 2022. https://www.watercalculator.org/news/news-briefs/renewable-power-95-percent-water-cut/.

9 John Locke, *Two Treatises of Government*, Everyman, 1993, London, England: Phoenix.

10 Ibid.

11 "Photosynthesis," *National Geographic*, January 22, 2024 https://education.nationalgeographic.org/resource/photosynthesis/.

12 "Soil Composition," University of Hawai'i at Manoa, 2023. https://www.ctahr.hawaii.edu/mauisoil/a_comp.aspx. Smithsonian National Museum of Natural History, *Dig It! The Secrets of Soil*. https://forces.si.edu/soils/04_00_13.html.

13 J. Gordon Betts et al., *Anatomy and Physiology*, Houston: Rice University, 2013, p. 43; Curt Stager, *Your Atomic Self*, p. 197.

14 "How Much Oxygen Comes from the Ocean?" NOAA's National Ocean Service. https://oceanservice.noaa.gov/facts/ocean-oxygen.html#:~:text=Scientists%20estimate%20that%20roughly%20half,smallest%20photosynthetic%20organism%20on%20Earth.

15 Graham P. Harris, *Phytoplankton Ecology: Structure, Function and Fluctuation*, London: Chapman & Hall, 1986; Yadigar Sekerci and Sergei Petrovskii, "Global Warming Can Lead to Depletion of Oxygen by Disrupting Phytoplankton Photosynthesis: A Mathematical Modelling Approach," *Geosciences* 8 (6), June 3, 2018. doi:10.3390/geosciences8060201.

16 Tony Allan. "The Virtual Water Concept." We World Energy, Water Stories, March 2020. https://www.eni.com/static/en-IT/world-energy-magazine/water-stories/We_WorldEnergy_46_eng.pdf.

17 Andrew Farmer, Samuela Bassi, and Malcolm Fergusson, "Water Scarcity and Droughts," European Parliament, February 2008: 35.

18 "Water Use: Virtual Water," Water Education Foundation. https://www.wateredducation.org/post/water-use-virtual-water; "Virtual Water," Econation, December 21, 2020. https://econation.one/virtual-water/; Thomas M. Kostigen, *The Green Blue Book: The Simple Water-Savings Guide to Everything in Your Life,* New York: Rodale Books, 2010.

19 Nicholas Kristof, "When One Almond Gulps 3.2 Gallons of Water," *New York Times*, May 13, 2023. https://www.nytimes.com/2023/05/13/opinion/water-shortage-west.html; Julian Fulton, Michael Norton, and Fraser Shilling, "Water-indexed Benefits and Impacts of California Almonds," *Ecological Indicators* 96 (1) 2019: 711–717. https://www.sciencedirect.com/science/article/pii/S1470160X17308592

20 Arjen Hoekstra and Ashok Chapagain. "Water Footprints of Nations: Water Use by People as a Function of Their Consumption Pattern," Integrated Assessment of Water Resources and Global Change, April 2007: 35–48. https://doi.org/10.1007/978-1-4020-5591-1_3.

21 Tony Allan, "The Virtual Water Concept". See note 16.

22 W. Z. Yang, L. Xu, Y. L. Zhao, L. Y. Chen, and T. A. McAllister. "Impact of Hard vs. Soft Wheat and Monensin Level on Rumen Acidosis in Feedlot Heifers." *Journal of Animal Science* 92 (11), November 2014: 5088-5098. doi: 10.2527/jas.2014-8092.

23 Anjuli Jain Figueroa, "How Much Water Did You Eat Today?" MIT J-WAFS, August 7, 2018. https://jwafs.mit.edu/news/2018/j-wafs-newsletter-highlight-how-much-water-did-you-eat-today.

24 Tony Allan, "The Virtual Water Concept". See note 16.

25 Ibid.

26 Mahima Shanker, "Virtual Water Trade." MAPL_1, May 25, 2022. https://www.maithriaqua.com/post/virtual-water-trade.

27 Erick Burgueño Salas, "Water Company Market Value Worldwide 2022," *Statista*, May 17, 2023. https://www.statista.com/statistics/1182423/leading-water-utilities-companies-by-market-value-worldwide/#:~:text=The%20water%20company%20with%20the,than%2029.4%20billion%20U.S.%20dollars.

28 Scott Lincicome, "Examining America's Farm Subsidy Problem," Cato Institute, December 18, 2020. https://www.cato.org/commentary/examining-americas-farm-subsidy-problem.

29 Tony Allan, "The Virtual Water Concept". See note 16.

30 "America Is Using Up Its Groundwater," *New York Times*. https://www.nytimes.com/interactive/2023/08/28/climate/groundwater-drying-climate-change.html?action=click&module=Well&pgtype=Homepage§ion=Climate%20and%20Environment.

31 Ibid.

32 Ibid.

33 Michele Thieme, "We Have Undervalued Freshwater; We Have Also Undervalued How Much It Matters," Deputy Director, WWF, October 16, 2023.

https://www.worldwildlife.org/blogs/sustainability-works/posts/we-have-undervalued-freshwater-we-have-also-undervalued-how-much-it-matters#:~:text=When%20considering%20the%20total%20footprint,%2C%20Japan%2C%20Germany%20and%20India.

34 Drew Swainston, "12 Drought-Tolerant Vegetables That Will Grow Well in Dry Conditions," June 1, 2023. https://www.homesandgardens.com/gardens/best-drought-tolerant-vegetables.

35 Tyler Ziton, "30 Best Drought-Tolerant Fruit and Nut Trees (Ranked)," Couch to Homestead, October 28, 2021. https://couchtohomestead.com/drought-tolerant-fruit-and-nut-trees/.

36 George Steinmetz, "A Five-Step Plan to Feed the World," Feeding 9 Billion – *National Geographic*. https://www.nationalgeographic.com/foodfeatures/feeding-9-billion/.

37 Will Henley, "Will We Ever See Water Footprint Labels on Consumer Products?" *Guardian*, August 23, 2013. https://www.theguardian.com/sustainable-business/water-footprint-labels-consumer-products#:~:text=Like%20Adeel%2C%20Davidoff%20believes%20water,go%20to%20measure%20water%20inputs.

38 Cynthia Larson, "Evidence of Shared Aspects of Complexity Science and Quantum Phenomena," *Cosmos and History: Journal of Natural and Social Philosophy* 12 (2), 2016.

39 David Wallace Wells, "Can We Put A Price on Climate Damages?" *New York Times*, September 20, 2023.

40 Abrahm Lustgarten and Meridith Kohut, "Climate Change Will Force a New American Migration," *ProPublica*, September 15, 2020. https://www.propublica.org/article/climate-change-will-force-a-new-american-migration.

41 Laura Lightbody and Brian Watts. "Repeatedly Flooded Properties Will Continue to Cost Taxpayers Billions of Dollars," The Pew Charitable Trusts, October 1, 2020. https://www.pewtrusts.org/en/research-and-analysis/articles/2020/10/01/repeatedly-flooded-properties-will-continue-to-cost-taxpayers-billions-of-dollars.

42 Abrahm Lustgarten and Meridith Kohut. "Climate Change Will Force a New American Migration." See note 40.

43 Christopher Flavelleer, Rick Rojas, Jim Tankersley, and Jack Healy, "Mississippi Crisis Highlights Climate Threat to Drinking Water Nationwide," *New York Times*, September 1, 2022. https://www.nytimes.com/2022/09/01/us/mississippi-water-climate-change.html?smid=nytcore-ios-share&referringSource=articleShare.

44 Ibid.

45 Ibid.

46 Ibid.

47 Ibid.

48 Aubri Juhasz, "Philadelphia Schools Close Due to High Temperatures and No Air Conditioning," NPR, August 31, 2022. https://www.npr.org/2022/08/31/1120355494/philadelphia-schools-close-due-to-high-temperatures-and-no-air-conditioning.

49 Ibid.

50 Ibid.

51 Ibid.

5 The Near Death and Rebirth of the Mediterranean

1 Patrick J. Kiger, "How Mesopotamia Became the Cradle of Civilization," *History*, November 10, 2020. https://www.history.com/news/how-mesopotamia-became-the-cradle-of-civilization.
2 Ibid.
3 N. S. Gill, "The Tigris River: Cradle of the Mesopotamian Civilization," *ThoughtCo.*, May 30, 2019. https://www.thoughtco.com/the-tigris-river-119231.
4 Thorkild Jacobsen and Robert M. Adams, "Salt and Silt in Ancient Mesopotamian Agriculture: Progressive Changes in Soil Salinity and Sedimentation Contributed to the Breakup of Past Civilizations," *Science* 128 (3334), November 21, 1958: 1251–1252.
5 Ibid.
6 Ibid.
7 Ibid.
8 Ibid.
9 Fred Pearce, *Keepers of the Spring: Reclaiming Our Water in an Age of Globalization.* Washington, DC: Island Press, 2004.
10 Eli Kotzer, "Artificial Kidneys for the Soil – Solving the Problem of Salinization of the Soil and Underground Water," *Desalination* 185 (2005): 71–77.
11 Shepard Krech, John Robert McNeill, and Carolyn Merchant, *Encyclopedia of World Environmental History*, New York: Routledge, 2004. pp. 1089–1090.
12 Jeremy Rifkin, *The Empathic Civilization: The Race to Global Consciousness in a World in Crisis*, New York: TarcherPerigee, 2009, p. 2.
13 IPCC, Rep. Climate Change 2021: The Physical Science Basis. Working Group I Contribution to the IPCC Sixth Assessment Report, Cambridge University Press, 2021. https://www.ipcc.ch/report/ar6/wg1/downloads/report/IPCC_AR6_WGI_SPM_final.pdf
14 "2022 State of Climate Services Energy," WMO, 2022. https://library.wmo.int/viewer/58116?medianame=1301_WMO_Climate_services_Energy_en_#page=2&viewer=picture&o=bookmarks&n=0&q=.
15 Ibid.
16 Josh Klemm and Isabella Winkler. "Which of the World's Hundreds of Thousands of Aging Dams Will Be the Next to Burst?" *New York Times*, September 17, 2023. https://www.nytimes.com/2023/09/17/opinion/libya-floods-dams.html.
17 WMO, "2022 State of Climate Services", 4. See note 14.
18 "The Mediterranean Eco-Region." NTPC DOCUMENT (NWFP FAO). https://www.fao.org/3/x5593e/x5593e02.htm#:~:text=This%20eco%2Dregion%20covers%20the,terrestrial%2C%20freshwater%20and%20marine%20ecosystems.
19 E. W. Ali, J. Cramer, E. Carnicer, N. Georgopoulou, G. Hilmi, Le Cozannet, and P. Lionello: Cross-Chapter Paper 4: Mediterranean Region. In: Climate Change 2022: Impacts, Adaptation and Vulnerability. Contribution of Working Group II to the Sixth Assessment Report of the Intergovernmental Panel on Climate Change, H.-O. Pörtner, D.C. Roberts, M. Tignor, E.S. Poloczanska, K. Mintenbeck, A. Alegría, M. Craig, S. Langsdorf, S. Löschke, V. Möller, A.

Okem, and B. Rama (eds.), Cambridge University Press, Cambridge, U.K. and New York, U.S.A., pp. 2233–2272. doi:10.1017/9781009325844.021. https://www.ipcc.ch/report/ar6/wg2/chapter/ccp4/.

20 "Climate Change in the Mediterranean," Climate change in the Mediterranean | UNEPMAP. https://www.unep.org/unepmap/resources/factsheets/climate-change#:~:text=The%20Mediterranean%20region%20is%20warming%2020%25%20faster%20than%20the%20global%20average.

21 Alexandre Tuel, Suchul Kang, and Elfatih A. Eltahir, "Understanding Climate Change over the Southwestern Mediterranean Using High-Resolution Simulations," *Climate Dynamics* 56 (3–4), November 2, 2020. 985–1001. https://doi.org/10.1007/s00382-020-05516-8.

22 "Climate Change in the Mediterranean." UNEPMAP. https://www.unep.org/unepmap/resources/factsheets/climate-change#:~:text=The%20Mediterranean%20region%20is%20warming%2020%25%20faster%20than%20the%20global%20average.

23 "Why the Mediterranean Is a Climate Change Hotspot." MIT Climate Portal. MIT News, June 17, 2020. https://climate.mit.edu/posts/why-mediterranean-climate-change-hotspot#:~:text=However%2C%20%E2%80%9CThere%20is%20one%20major,of%20any%20landmass%20on%20Earth.

24 Samya Kullab, "Politics, Climate Conspire as Tigris and Euphrates Dwindle," *AP NEWS*, Associated Press, November 18, 2022. https://apnews.com/article/iran-middle-east-business-world-news-syria-3b8569a74d798b9923e2a8b812fa1fca.

25 Hamza Ozguler and Dursun Yildiz, "Consequences of the Droughts in the Euphrates– Tigris Basis," *Water Management and Diplomacy* 1, 2020. https://dergipark.org.tr/tr/download/article-file/1151377

26 Tomer Barak and Hay Eytan Cohen Yanarocak, "Confronting Climate Change, Turkey Needs 'Green' Leadership Now More than Ever," Middle East Institute, January 25, 2022. https://www.mei.edu/publications/confronting-climate-change-turkey-needs-green-leadership-now-more-ever.

27 M. Türkeş, "İklim Verileri Kullanılarak Türkiye'nin Çölleşme Haritası Dokümanı Hazırlanması Raporu," Orman ve Su, "İşleri Bakanlığı, Çölleşme ve Erozyonla Mücadele Genel Müdürlüğü Yayını," Ankara, Turkey, 2013, p. 57. https://www.researchgate.net/publication/293334692_Iklim_Verileri_Kullanilarak_Turkiye'nin_Collesme_Haritasi_Dokumani_Hazirlanmasi_Raporu.

28 Caterina Scaramelli, "The Lost Wetlands of Turkey," *MERIP*, October 20, 2020. https://merip.org/2020/10/the-lost-wetlands-of-turkey/.

29 Abbie Cheeseman, "Iraq's Mighty Rivers Tigris and Euphrates 'Will Soon Run Dry," *The Times*, December 3, 2021. https://www.thetimes.co.uk/article/iraqs-mighty-rivers-tigris-and-euphrates-will-soon-run-dry-q5h72g5sk.

30 Ibid.

31 Samya Kullab, "Politics, Climate Conspire as Tigris and Euphrates Dwindle," Associated Press, November 18, 2022. https://apnews.com/article/iran-middle-east-business-world-news-syria-3b8569a74d798b9923e2a8b812fa1fca.

32 Ibid.

33 "Migration, Environment, and Climate Change in Iraq," United Nations. https://iraq.un.org/en/194355-migration-environment-and-climate-change-iraq.

34 "A 3,400-Year-Old City Emerges from the Tigris River," University of Tübingen, February 2, 2023. https://uni-tuebingen.de/en/university/news-and-publications/press-releases/press-releases/article/a-3400-year-old-city-emerges-from-the-tigris-river/.

35 Paul Hockenos, "As the Climate Bakes, Turkey Faces a Future without Water," Yale E360, September 30, 2021. https://e360.yale.edu/features/as-the-climate-bakes-turkey-faces-a-future-without-water.

36 Ibid.

37 Ibid.

38 Ibid.

39 Ercan Ayboga, "Policy and Impacts of Dams in the Euphrates and Tigris Basins," Paper for the Mesopotamia Water Forum 2019, Sulaymaniyah, Kurdistan Region of Iraq, 2. https://www.savethetigris.org/wp-content/uploads/2019/01/Paper-Challenge-B-Dams-FINAL-to-be-published.pdf; Ercan Ayboga and I. Akgun, "Iran's Dam Policy and the Case of Lake Urmia,' 2012. www.ekopotamya.net/index.php/2012/07/irans-dam-policy-and-the-case-of-the-lake-urmia/; K. Madani, "Water Management in Iran: What Is Causing the Looming Crisis?" *Journal of Environmental Studies and Sciences*, 4 (4), 2014: 315–328.

40 "10 Years on, Turkey Continues Its Support for an Ever-Growing Number of Syrian Refugees," World Bank, June 22, 2021. https://www.worldbank.org/en/news/feature/2021/06/22/10-years-on-turkey-continues-its-support-for-an-ever-growing-number-of-syrian-refugees; "Climate Change, War, Displacement, and Health: The Impact on Syrian Refugee Camps – Syrian Arab Republic," *ReliefWeb*, September 20, 2022. https://reliefweb.int/report/syrian-arab-republic/climate-change-war-displacement-and-health-impact-syrian-refugee-camps.

41 "Istanbul Population 2023." https://worldpopulationreview.com/world-cities/istanbul-population.

42 Akgün İlhan, "Istanbul's Water Crisis," *Green European Journal*, November 8, 2021. https://www.greeneuropeanjournal.eu/istanbuls-water-crisis/.

43 Katy Dartford, "Turkey Faces Its Most Severe Drought in a Decade," euronews, January 14, 2021, https://www.euronews.com/my-europe/2021/01/14/pray-for-rain-ceremonies-are-useless-turkey-faces-its-most-severe-drought-in-a-decade.

44 Akgün İlhan, "Istanbul's Water Crisis." See note 42.

45 Dave Chambers, "Icebergs to Save Cape Town from Drought Would Be Drop in the Ocean," News24, January 11, 2023. https://www.news24.com/news24/bi-archive/icebergs-to-save-cape-town-from-drought-would-be-drop-in-the-ocean-2023-1#; Alan Condron, "Towing Icebergs to Arid Regions to Reduce Water Scarcity," *Scientific Reports* 13 (1), January 7, 2023. https://doi.org/10.1038/s41598-022-26952-y.

46 William Hale, "Turkey's Energy Dilemmas: Changes and Challenges," *Middle Eastern Studies* 58 (3), 2022: 453. DOI: 10.1080/00263206.2022.2048478.

47 Rep. *Turkey 2021: Energy Policy Review*. International Energy Agency, March 2021. https://iea.blob.core.windows.net/assets/cc499a7b-b72a-466c-88de-d792a9daff44/Turkey_2021_Energy_Policy_Review.pdf.

48 William Hale, "Turkey's Energy Dilemmas." See note 46.

49 "Renewable Power's Growth is being Turbocharged as Countries Seek to Strengthen Energy Security," International Energy Agency, 6 December 2022. https://www.iea.org/news/renewable-power-s-growth-is-being-turbocharged-as-countries-seek-to-strengthen-energy-security
50 Gareth Chetwynd, "Spain Eyes Massive Solar and Wind Boosts under New Energy Plan," *Recharge*, June 29, 2023. https://www.rechargenews.com/energy-transition/spain-eyes-massive-solar-and-wind-boosts-under-new-energy-plan/2-1-1477558.
51 Ibid.
52 Shaheena Uddin, news reporter. "For Five Hours Last Week Greece Ran Entirely on Electricity from Solar, Wind and Water," *Sky News*, October 14, 2022, https://news.sky.com/story/for-five-hours-last-week-greece-ran-entirely-on-electricity-from-solar-wind-and-water-12720353#:~:text=Greece%20ran%20entirely%20on%20renewable,country's%20independent%20power%20transmission%20operator.
53 Monica Tyler Davies, "A New Fossil Free Milestone: $11 Trillion Has Been Committed to Divest from Fossil Fuels," 350 Action, September 11, 2019. https://350.org/11-trillion-divested/.
54 International Energy Agency, October 27, 2022. https://www.iea.org/news/world-energy-outlook-2022-shows-the-global-energy-crisis-can-be-a-historic-turning-point-towards-a-cleaner-and-more-secure-future.
55 William Hale, "Turkey's Energy Dilemmas." See note 46.
56 Ibid.
57 Elena Ambrosetti, "Demographic Challenges in the Mediterranean, Panorama." https://www.iemed.org/wp-content/uploads/2021/01/Demographic-Challenges-in-the-Mediterranean.pdf.
58 Turkey: Energy Policy Review. International Energy Agency, p. 78. https://iea.blob.core.windows.net/assets/cc499a7b-b72a-466c-88de-d792a9daff44/Turkey_2021_Energy_Policy_Review.pdf.
59 "Turkey Green Energy and Clean Technologies," International Trade Administration, April 22, 2022. https://www.trade.gov/market-intelligence/turkey-green-energy-and-clean-technologies.
60 "Turkey's Installed Solar Power Capacity to Exceed 30 GW by 2030," *Daily Sabah*, 20 June 2022, https://www.dailysabah.com/business/energy/turkeys-installed-solar-power-capacity-to-exceed-30-gw-by-2030.
61 Burhan Yuksekkas, "Turkish Companies Go Solar at Record Pace to Cut Energy Costs," Bloomberg, December 1, 2022. https://www.bloomberg.com/news/articles/2022-12-01/turkey-solar-panel-demand-booms-as-companies-avoid-rising-power-costs.
62 Ibid.
63 "Turkey's Installed Solar Power Capacity." See note 60.
64 A. J. Dellinger, "Gigawatt: The Solar Energy Term You Should Know About," CNET, November 16, 2021. https://www.cnet.com/home/energy-and-utilities/gigawatt-the-solar-energy-term-you-should-know-about/.
65 Joyce Lee and Feng Zhao, "Global Wind Report 2022," Global Wind Energy Council, April 4, 2022, p. 138. https://gwec.net/wp-content/uploads/2022/04/Annual-Wind-Report-2022_screen_final_April.pdf.
66 "Turkey Reaches 10 GW Wind Energy Milestone," Wind Europe, September

9, 2021. https://windeurope.org/newsroom/news/turkey-reaches-10-gw-wind-energy-milestone/.

67 Alfredo Parres, "Grid Integration Key to Turkey's Wind Power Success," ABB Conversations, March 30, 2015. https://www.abb-conversations.com/2015/03/grid-integration-key-to-turkeys-wind-power-success/.

68 "Turkey Holds 75 Gigawatts of Offshore Wind Energy Potential," *Daily Sabah*, April 19, 2021. https://www.dailysabah.com/business/energy/turkey-holds-75-gigawatts-of-offshore-wind-energy-potential; and see note 60; Eylem Yilmaz Ulu and Omer Altan Dombayci, "Wind Energy in Turkey: Potential and Development," *Eurasia Proceedings of Science, Technology, Engineering, and Mathematics* 4, 2018: 132–136. http://www.epstem.net/tr/download/article-file/595454.

69 "Turkey Reaches 10 GW Wind Energy Milestone," Wind Europe, September 9, 2021. https://windeurope.org/newsroom/news/turkey-reaches-10-gw-wind-energy-milestone/.

70 Takvor Soukissian, Flora E. Karathanasi, and Dimitrios K. Zaragkas, "Exploiting Offshore Wind and Solar Resources in the Mediterranean Using ERA5 Reanalysis Data, 2021." https://arxiv.org/pdf/2104.00571.pdf.

71 "The Hydrogen Colour Spectrum," National Grid Group. https://www.nationalgrid.com/stories/energy-explained/hydrogen-colour-spectrum#:~:text=Grey%20hydrogen,gases%20made%20in%20the%20process; Catherine Clifford, "Hydrogen Power is Gaining Momentum, but Critics Say it's neither Efficient nor Green Enough," CNBC, January 6, 2022. https://www.cnbc.com/2022/01/06/what-is-green-hydrogen-vs-blue-hydrogen-and-why-it-matters.html.

72 Turner Jackson, "3 Questions: Blue Hydrogen and the World's Energy Systems," *MIT News*, Massachusetts Institute of Technology, MIT Energy Initiative, October 17, 2022. https://news.mit.edu/2022/3-questions-emre-gencer-blue-hydrogen-1017#:~:text=hydrogen%20production%20processes.-,Natural%20gas%2Dbased%20hydrogen%20production%20with%20carbon%20capture%20and%20storage,a%20low%2Dcarbon%20energy%20carrier.; Marsh, Jane, "Hydrogen for Clean Energy could Be Produced from Seawater," *Sustainability Times*, October 19, 2022, https://www.sustainability-times.com/low-carbon-energy/hydrogen-for-clean-energy-could-be-produced-from-seawater/; Shawn Johnson, "Water-Splitting Device Solves Puzzle of Producing Hydrogen Directly from Seawater," *BusinessNews*, December 6, 2022. https://biz.crast.net/water-splitting-device-solves-puzzle-of-producing-hydrogen-directly-from-seawater/; Yun Kuang et al. "Solar-Driven, Highly Sustained Splitting of Seawater into Hydrogen and Oxygen Fuels," *Proceedings of the National Academy of Science* 116 (14), April 2, 2019, https://www.pnas.org/doi/10.1073/pnas.1900556116#bibliography.

73 Darius Snieckus, "World's Largest Floating Wind-Fueled H2 Hub in Frame for Italian Deepwater 'by 2027,'" *Recharge*, September 26, 2022. https://www.rechargenews.com/energy-transition/worlds-largest-floating-wind-fuelled-h2-hub-in-frame-for-italian-deepwater-by-2027/2-1-1320795.

74 "The Precautionary Principle," *Eur-Lex*, November 30, 2016. https://eur-lex.europa.eu/EN/legal-content/summary/the-precautionary-principle.html.

75 Ibid.

76 Martina Bocci and Francesca Coccon, "Using Ecological Sensitivity to Guide Marine Renewable Energy Potentials in the Mediterranean Region," Interreg Mediterranean Fact Sheet, 2020, p. 1. https://planbleu.org/wp-content/uploads/2021/03/MBPC_Technical_Factsheet_on_BAT___BEP_for_Marine_Renewable_Energy_FINAL.pdf.
77 Ibid., 12.
78 Ibid., 14.
79 "Renewable Energy – Powering a Safer Future," United Nations. https://www.un.org/en/climatechange/raising-ambition/renewable-energy.
80 Antonio Zapata-Sierra et al., "Worldwide Research Trends on Desalination," *Desalination* 519, 2022. https://www.sciencedirect.com/science/article/pii/S0011916421003763.
81 Ibid., 1.
82 John Tonner, "Barriers to Thermal Desalination in the United States," Desalination and Water Purification Research and Development Program Report No. 144, U.S. Department of the Interior Bureau of Reclamation, March 2008. https://www.usbr.gov/research/dwpr/reportpdfs/report144.pdf.
83 Hesham R. Lofty et al., "Renewable Energy Powered Membrane Desalination – Review of Recent Development," *Environmental Science and Pollution Research* 29, 2022. https://link.springer.com/article/10.1007/s11356-022-20480-y.
84 Abdul Latif Jameel, "Fresh Water; Fresh Ideas. Can Renewable Energy be the Future of Desalination?," November 16, 2020, https://alj.com/en/perspective/fresh-water-fresh-ideas-can-renewable-energy-be-the-future-of-desalination/; Laura F. Zarza, "Spanish Desalination Know-How, a Worldwide Benchmark," *Smart Water Magazine*, February 28, 2022. https://smartwatermagazine.com/news/smart-water-magazine/spanish-desalination-know-how-a-worldwide-benchmark.
85 Hesham R. Lofty et al., "Renewable Energy Powered Membrane Desalination." See note 83.
86 Molly Walton, "Desalinated Water Affects the Energy Equation in the Middle East," International Energy Agency, 21 January 2019, https://www.iea.org/commentaries/desalinated-water-affects-the-energy-equation-in-the-middle-east.
87 "Water Desalination Using Renewable Energy," IEA-ETSAP and IRENA Technology Brief, 12 March 2012, Pg 1, https://www.irena.org/-/media/Files/IRENA/Agency/Publication/2012/IRENA-ETSAP-Tech-Brief-I12-Water-Desalination.pdf.
88 "Global Clean Water Desalination Alliance (GCWDA)." Global Clean Water Desalination Alliance (GCWDA) – Climate Initiatives Platform. https://climateinitiativesplatform.org/index.php/Global_Clean_Water_Desalination_Alliance_(GCWDA).
89 Abdul Latif Jameel, "Fresh Water; Fresh Ideas." See note 84; "The Role of Desalination in an Increasingly Water-Scarce World," World Bank Group, 2019, p. 57, https://documents1.worldbank.org/curated/en/476041552622967264/pdf/135312-WP-PUBLIC-14-3-2019-12-3-35-W.pdf.
90 Abdul Latif Jameel, "Fresh Water; Fresh Ideas." See note 84.
91 Aidan Lewis, "Egypt to Build 21 Desalination Plants in Phase 1 of Scheme – Sovereign Fund," Reuters, 1 December 2022, https://www.reuters.com/markets/commodities/egypt-build-21-desalination-plants-phase-1-scheme-sovereign-fund-2022-12-01/.

92 "ACCIONA Starts Construction of Jubail 3B Desalination Plant in Saudi Arabia," ACCIONA press release, June 9, 2022. https://www.acciona.com/updates/articles/acciona-starts-construction-jubail-3b-desalination-plant-saudi-arabia/?_adin=02021864894.

93 Susan Kraemer, "Australia Gets Ten Times Bigger Solar Farm Following Carbon Tax," *CleanTechnica*, September 2, 2011. https://cleantechnica.com/2011/09/01/australia-gets-ten-times-bigger-solar-farm-following-carbon-tax/.

94 Simon Atkinson, "Precisely Controlling the Density of Water Filtration Membranes Increases Their Efficiency, Shows Research." Membrane Technology 8, December 11, 2021: 5–6. https://doi.org/10.1016/s0958-2118(21)00124-5.

95 David L. Chandler, "Turning Desalination Waste into a Useful Resource," MIT, May 15, 2019. https://energy.mit.edu/news/turning-desalination-waste-into-a-useful-resource/.

96 Daniel Hickman and Raffaele Molinari, "Can Brine from Seawater Desalination Plants Be a Source of Critical Metals?" *ChemistryViews*, September 25, 2023. https://www.chemistryviews.org/details/ezine/11347408/can_brine_from_seawater_desalination_plants_be_a_source_of_critical_metals/.

97 Robert Strohmeyer, "The 7 Worst Tech Predictions of All Time," *PCWorld*, December 31, 2008. https://www.pcworld.com/article/532605/worst_tech_predictions.html.

98 Peter J. Denning and Ted G. Lewis, "Exponential Laws of Computing Growth," *Communications of the ACM* 60 (1), January 2017. https://cacm.acm.org/magazines/2017/1/211094-exponential-laws-of-computing-growth/abstract

99 Petroc Taylor, "Smartphone Subscriptions Worldwide 2016–2021, with forecasts from 2022 to 2027," *Statista*, July 19, 2023. https://www.statista.com/statistics/330695/number-of-smartphone-users-worldwide/.

100 Wafa Suwaileh, Daniel Johnson, and Nidal Hilal, "Membrane Desalination and Water Re-Use for Agriculture: State of the Art and Future Outlook," *Desalination* 491, October 1, 2020. https://www.sciencedirect.com/science/article/abs/pii/S0011916420310213.

6 Location, Location, Location: The Eurasian Pangaea

1 "China–EU – International Trade in Goods Statistics," Statistics Explained, February 2022. https://ec.europa.eu/eurostat/statistics-explained/index.php?title=China-EU_-_international_trade_in_goods_statistics#:~:text=China%20largest%20partner%20for%20EU%20imports%20of%20goods%20in%202022,-The%20position%20of&text=It%20was%20the%20largest%20partner,and%20Norway%20(5.4%20%25).

2 James McBride et al., "China's Massive Belt and Road Initiative, Council on Foreign Relations, 2023. https://www.cfr.org/backgrounder/chinas-massive-belt-and-road-initiative.

3 "About the Belt and Road Initiative (BRI)," Green Finance & Development Center. https://greenfdc.org/belt-and-road-initiative-about/.

4 Suprabha Baniya, Nadia Rocha, and Michele Ruta, "Trade Effects of the New Silk Road: A Gravity Analysis," World Bank Policy Research Working Paper 8694, January 2019; Michele Ruta et al., "How much will the Belt and Road Initiative Reduce Trade Costs?" World Bank, October 16, 2018. https://blogs.worldbank.org/trade/how-much-will-belt-and-road-initiative-reduce-

trade-costs. "Belt and Road Initiative to boost world GDP by over $7 trillion per annum by 2040," Centre for Economics and Business Research, May 27, 2019, https://cebr.com/reports/belt-and-road-initiative-to-boost-world-gdp-by-over-7-trillion-per-annum-by-2040/.

5 Ibid.

6 Nicolas J. Firzli, "Pension Investment in Infrastructure Debt: A New Source of Capital for Project Finance," World Bank, May 24, 2016. https://blogs.worldbank.org/ppps/pension-investment-infrastructure-debt-new-source-capital-project-finance.

7 Charlie Campbell, "China Says It's Building the New Silk Road. Here Are Five Things to Know Ahead of a Key Summit," *Time*, May 12, 2017. https://time.com/4776845/china-xi-jinping-belt-road-initiative-obor/; James Griffiths, " Just what is this One Belt, One Road thing anyway?" CNN, May 11, 2017. https://www.cnn.com/2017/05/11/asia/china-one-belt-one-road-explainer/index.html.

8 Felix K. Chang, "The Middle Corridor through Central Asia: Trade and Influence Ambitions," Foreign Policy Research Institute, February 21, 2023. https://www.fpri.org/article/2023/02/the-middle-corridor-through-central-asia-trade-and-influence-ambitions/.

9 Laura, Basagni, "The Mediterranean Sea and its Port System: Risk and Opportunities in a Globally Connected World," p. 13, German Marshall Fund. https://www.gmfus.org/sites/default/files/Chapter%20Laura%20Basagni__JPS_Infrastructures%20and%20power%20in%20the%20MENA-12-33.pdf.

10 Ibid., 13.

11 Ibid.

12 Michele Barbero, "Europe Is Trying (and Failing) to Beat China at the Development Game," *Foreign Policy*, Graham Digital Holding Company, January 10, 2023. https://foreignpolicy.com/2023/01/10/europe-china-eu-global-gateway-bri-economic-development/; "Demographic Change + Export Controls + Global Gateway." Merics, February 2, 2023. https://merics.org/en/merics-briefs/demographic-change-export-controls-global-gateway.

13 "Bioregion," European Environment Agency. https://www.eea.europa.eu/help/glossary/chm-biodiversity/bioregion.

14 "EuroRegion." euroregion.edu, 2021. https://euroregio.eu/en/euroregion.

15 Programmes of the Catalan Presidency of a Euroregion Pyrenees – 2023–2025

16 "The Future Looks Bright for Solar Energy in Jordan: A 2023 Outlook," *SolarQuarter*, February 25, 2023. https://solarquarter.com/2023/02/25/the-future-looks-bright-for-solar-energy-in-jordan-a-2023-outlook/#:~:text=According%20to%20a%20report%20by,reliance%20on%20imported%20fossil%20fuels.

17 "Green Blue Deal," *EcoPeace Middle East*, March 31, 2022. https://ecopeaceme.org/gbd/.

7 Freeing the Waters

1 Elizabeth Pennisi, "Just 19% of Earth's Land Is Still 'Wild,' Analysis Suggests" *Science*, April 19, 2021. https://www.science.org/content/article/just-19-earths-land-still-wild-analysis-suggests.

2 Michelle Nijhuis, "World's Largest Dam Removal Unleashes U.S. River After Century of Electric Production," *National Geographic*, May 4, 2021. https://www.nationalgeographic.com/science/article/140826-elwha-river-dam-removal-salmon-science-olympic.
3 Sarah Laskow, "Finding Brooklyn's Ghost Streams, with Old Maps and New Technology," *Atlas Obscura*, January 8, 2016. https://www.atlasobscura.com/articles/finding-brooklyns-ghost-streams-with-old-maps-and-new-technology.
4 Adam Shell, "No U.S. Stock, Bond Trading Monday, Tuesday," *USA Today*, October 29, 2012. https://www.usatoday.com/story/money/markets/2012/10/28/nyse-sandy/1664249/; "Impact of Hurricane Sandy." https://www.nyc.gov/html/sirr/downloads/pdf/final_report/Ch_1_SandyImpacts_FINAL_singles.pdf
5 Fran Southgate, "Rewilding Water," *Rewilding Britain*. https://www.rewildingbritain.org.uk/why-rewild/what-is-rewilding/examples/rewilding-water.
6 Ibid.
7 Ibid.
8 Ibid.
9 Joshua Larsen and Annegret Larsen, "Rewilding: Beavers Are Back – Here's What This Might Mean for the U.K." *Positive News*, September 24, 2021. https://www.positive.news/environment/rewilding-beavers-are-back-heres-what-this-might-mean/.
10 Marvin S. Soroos, "The International Commons: A Historical Perspective," *Environmental Review* 12 (1), Spring 1988: 1–22, https://www.jstor.org/stable/3984374.
11 Sir Walter Raleigh, "A Discourse of the Invention of Ships, Anchors, Compass, & etc.," in *Oxford Essential Quotations*, ed. Susan Racliffe, 2017. https://www.oxfordreference.com/view/10.1093/acref/9780191843730.001.0001/qoroed500008718.
12 William E. Livezey, *Mahan on Sea Power*, Norman: University of Oklahoma Press, 1981, pp. 281–282, https://www.baltdefcol.org/files/files/BSDR/BSDR_11_2.pdf.
13 Clive Schofield and Victor Prescott, *The Maritime Political Boundaries of the World*, Leiden: Martinus Nijhoff, 2004, p. 36; Food and Agriculture Organization of the United Nations, "The State of World Fisheries and Aquaculture 2020. Sustainability in Action," 2020, 94; "United Nations Convention on the Law of the Sea (UNCLOS)." Environmental Science: In Context. *Encyclopedia.com*. January 8, 2024. https://www.encyclopedia.com/environment/energy-government-and-defense-magazines/united-nations-convention-law-sea-unclos.; "Opposition to New Offshore Drilling in the Pacific Ocean," *Oceana USA*, August 29, 2022. https://usa.oceana.org/pacific-drilling/.
14 "Overfishing in the Georges Bank: AMNH." American Museum of Natural History, 2013. https://www.amnh.org/explore/videos/biodiversity/georges-bank-fish-restoration.
15 Ibid.
16 Alison Chase, "Marine Protected Areas Are Key to Our Future," *Natural Resources Defense Council*, June 14, 2021. https://www.nrdc.org/bio/alison-chase/marine-protected-areas-are-key-our-future#:~:text=Fully%20and%20highly%20protected%20marine,and%20the%20jobs%20they%20generate.

17 Matt Rand, "Study Shows Benefits Extend beyond Sea Life to Communities on Land," The Pew Charitable Trusts, July 7, 2020. https://www.pewtrusts.org/en/research-and-analysis/articles/2020/07/07/marine-reserves-can-help-oceans-and-people-withstand-climate-change.

18 David Stanway, "Nations Secure U.N. Global High Seas Biodiversity Pact," Reuters, March 6, 2023. https://www.reuters.com/business/environment/nations-secure-un-global-high-seas-biodiversity-pact-2023-03-05/.

19 Kevin McAdam, "The Human Right to Water – Market Allocations and Subsistence in a World of Scarcity," *The Interdisciplinary Journal of Study Abroad*, 2003: 59–85. https://doi.org/https://files.eric.ed.gov/fulltext/EJ891474.pdf.

20 Erick Burgueño Salas, "Water Company Market Value Worldwide 2022," Statista, May 17, 2023. https://www.statista.com/statistics/1182423/leading-water-utilities-companies-by-market-value-worldwide/#:~:text=The%20water%20company%20with%20the,electricity%20and%20natural%20gas%20services.

21 "Water Privatization: Facts and Figures," Food & Water Watch, March 29, 2023. https://www.foodandwaterwatch.org/2015/08/02/water-privatization-facts-and-figures/.

22 Bobby Magill, "Climate Change Could Increase Global Fresh Water," MIT, Climate Central, October 2, 2014. https://www.climatecentral.org/news/climate-change-could-increase-global-fresh-water-supply-mit-18124; "2014 Energy and Climate Outlook," MIT Joint Program on the Science and Policy of Global Change," 2014. https://globalchange.mit.edu/sites/default/files/newsletters/files/2014%20Energy%20%26%20Climate%20Outlook.pdf.

23 "2014 Energy and Climate Outlook". See Ibid.

24 Bobby Magill, "Climate Change Could Increase Global Fresh Water." See note 22.

25 Ibid.

26 Erica Gies, "Slow Water: Can We Tame Urban Floods by Going with the Flow?" *Guardian*. Guardian News and Media, June 7, 2022. https://www.theguardian.com/environment/2022/jun/07/slow-water-urban-floods-drought-china-sponge-cities.

27 Ibid.

28 "Stormwater Tip: How are Bioswales and Rain Gardens Different?". Pittsburgh Water & Sewer Authority. June 2021 https://www.pgh2o.com/news-events/news/newsletter/2021-06-29-stormwater-tip-how-are-bioswales-and-rain-gardens-different.

29 "Using Green Roofs to Reduce Heat Islands," Environmental Protection Agency (EPA). https://www.epa.gov/heatislands/using-green-roofs-reduce-heat-islands#1.

30 Stefano Salata and Bertan Arslan, "Designing with Ecosystem Modelling: The Sponge District Application in Izmir, Turkey," *Sustainability* 14 2022: 3420. https:// doi.org/10.3390/su14063420

31 Jared Green, "Kongjian Yu Defends His Sponge City Campaign," *The Dirt*, August 4, 2021. https://dirt.asla.org/2021/08/04/kongjian-yu-defends-his-sponge-city-campaign/.

32 Brad Lancaster, "Roman- and Byzantine-Era Cisterns of the Past Reviving Life in the Present," in *Rainwater Harvesting for Drylands and Beyond*, 2011. https://www.harvestingrainwater.com/2011/07/roman-and-byzantine-era-cisterns-of-the-past-reviving-life-in-the-present/.

33 "Rainwater Conservation for Community Climate Change Resiliency," March

5, 2020. https://www.peacecorps.gov/mexico/stories/rainwater-conservation-community-climate-change-resiliency/.

34 "One Million Cisterns for the Sahel Initiative," 2018. https://www.fao.org/3/ca0882en/CA0882EN.pdf.

35 Ibid., 1.

36 "President of Niger: 'Development Is the Only Way to Stop Migration,'" FAO, June 19, 2018. https://www.fao.org/news/story/en/item/1141812/icode/.

37 Alexander Otte. "Chapter 3: Social Dimensions." Leaving No One Behind, The United Nations World Water Development Report, UNESCO, 2019. https://unesdoc.unesco.org/ark:/48223/pf0000367652.

38 Rainwater Collection Legal States 2024, 2024. https://worldpopulationreview.com/state-rankings/rainwater-collection-legal-states.

39 Harriet Festing et al., "The Case for Fixing the Leaks: Protecting People and Saving Water while Supporting Economic Growth in the Great Lakes Region," Center for Neighborhood Technology, 2013. https://cnt.org/sites/default/files/publications/CNT_CaseforFixingtheLeaks.pdf.

40 Bob Berkebile et al., "Flow – The Making of the Omega Center for Sustainable Living," BNIM, 2010. https://www.bnim.com/sites/default/files/library/flow_0.pdf.; "The Eco Machine." Omega Institute for Holistic Studies, 2023. https://www.eomega.org/center-sustainable-living/eco-machine.

41 Ibid.

42 Jim Robbins, "Beyond the Yuck Factor: Cities Turn to Extreme Water Recycling," Yale Environment 360, June 6, 2023. https://e360.yale.edu/features/on-site-distributed-premise-graywater-blackwater-recycling.

43 Ibid.

44 E. Pinkham and M. Woodson, "Salesforce announces Work.com for schools and $20 million to help schools reopen safely and Support Student Learning Anywhere." Salesforce. August 11, 2020. https://www.salesforce.com/news/press-releases/2020/08/11/salesforce-announces-work-com-for-schools-and-20-million-to-help-schools-reopen-safely-and-support-student-learning-anywhere/.

45 Patrick Sisson. "Facing Severe Droughts, Developers Seek to Reuse the Water They Have," *The New York Times*, August 3, 2021. https://www.nytimes.com/2021/08/03/business/drought-water-reuse-development.html.

46 Jim Robbins, "Beyond the Yuck Factor: Cities Turn to Extreme Water Recycling," Yale Environment 360, June 6, 2023. https://e360.yale.edu/features/on-site-distributed-premise-graywater-blackwater-recycling.

47 Ibid.

48 Ibid.

49 Ibid.

8 The Great Migration and the Rise of Ephemeral Society

1 Abrahm Lustgarten and Meridith Kohut, "Climate Change Will Force a New American Migration," *ProPublica*, September 15, 2020. https://www.propublica.org/article/climate-change-will-force-a-new-american-migration.

2 Lee R. Kump, "The Last Great Global Warming," *Scientific American*, July 1, 2011. https://www.scientificamerican.com/article/the-last-great-global-warming/.

3 Abbey of Regina Laudis: St. Benedict's rule. https://abbeyofreginalaudis.org/community-rule-english.html.

4 Sebastian de Grazia, *Of Time, Work, and Leisure*, New York: Century Foundation, 1962, p. 41.
5 Ibid.
6 Reinhard Bendix and Max Weber, *An Intellectual Portrait*, Garden City: Anchor-Doubleday, 1962, p. 318.
7 Jonathan Swift, 1667–1745, *Gulliver's Travels*, New York, Avenel Books, 1985.
8 "Linear Perspective," *Encyclopedia Britannica*. https://www.britannica.com/art/linear-perspective.
9 Fritjof Capra, *The Tao of Physics: An Exploration of the Parallels Between Modern Physics and Eastern Mysticism*, Berkeley: Shambhala Publications, 1975, p. 138.
10 Norbert Wiener, *The Human Use of Human Beings: Cybernetics and Society*, New York: Da Capo Press, 1988, p. 96.
11 Alfred North Whitehead, *Science and the Modern World*, Cambridge University Press, 1926, p. 22.
12 Alfred North Whitehead, *Science and the Modern World*: Lowell Lectures1925, Cambridge University Press, 1929, p. 61; Alfred North Whitehead, *Nature and Life*, Chicago University Press, 1934, and reprinted Cambridge University Press, 2011.
13 Alfred North Whitehead. See two works, note 12.
14 Whitehead, *Nature and Life*, p. 65.
15 Robin G. Collingwood, *The Idea of Nature*, Oxford University Press, 1945, p. 146.
16 Whitehead, *Nature and Life*, pp. 45–48.
17 Allan Silverman, "Plato's Middle Period Metaphysics and Epistemology," *Stanford Encyclopedia of Philosophy*, ed. Edward N. Zalta, Fall 2014 edn. https://plato.stanford.edu/archives/fall2014/entries/plato-metaphysics
18 Vernon J. Bourke, "Rationalism." In *Dictionary of Philosophy*, ed. Dagobert D. Runes, 263. Totowa, NJ: Littlefield, Adams, and Company, 1962.
19 Isaac Newton, 1642–1727, *Newton's Principia: The Mathematical Principles of Natural Philosophy*, New York: Daniel Adee, 1846.
20 "Ephemeral Art" *UNESCO Courier*, 1996, p. 11. https://unesdoc.unesco.org/ark:/48223/pf0000104975.
21 "Annual Park Ranking Report for Recreation Visits in 2022," National Parks Service. https://irma.nps.gov/Stats/SSRSReports/National%20Reports/Annual%20Park%20Ranking%20Report%20(1979%20-%20Last%20Calendar%20Year.
22 Mary Caperton Morton, "Mount Rushmore's Six Grandfathers and Four Presidents," *Eos*, October 14, 2021. https://eos.org/features/mount-rushmores-six-grandfathers-and-four-presidents; Mario Gonzalez and Elizabeth Cook-Lynn, *The Politics of Hallowed Ground: Wounded Knee and the Struggle for Indian Sovereignty*, Urbana: University of Illinois Press, 1999.
23 Peter Osborne and Matthew Charles, "Walter Benjamin," *Stanford Encyclopedia of Philosophy*, October 14, 2020. https://plato.stanford.edu/entries/benjamin/.

9 Rethinking Attachment to Place: Where We've Come from and Where We're Heading

1 K. C. Samir, and Wolfgang Lutz. "The Human Core of the Shared Socioeconomic Pathways: Population Scenarios by Age, Sex and Level of Education for All Countries to 2100," *Global Environmental Change*, July 4, 2014. https://www.sciencedirect.com/science/article/pii/S0959378014001095.

2 Dean Spears, "All of the Predictions Agree on One Thing: Humanity Peaks Soon," *New York Times*, September 18, 2023. https://www.nytimes.com/interactive/2023/09/18/opinion/human-population-global-growth.html.

3 "The Great Human Migration," Smithsonian.com, July 1, 2008. https://www.smithsonianmag.com/history/the-great-human-migration-13561/.

4 Patrick Manning and Tiffany Trimmer, *Migration in World History*, New York: Routledge, 2020.

5 Ibid., 33.

6 K. R. Howe, *Vaka Moana: Voyages of the Ancestors: The Discovery and Settlement of the Pacific*, Honolulu: University of Hawai'i Press, 2014.

7 Kim Tingley, "The Secrets of the Wave Pilots," *New York Times*, March 17, 2016, https://www.nytimes.com/2016/03/20/magazine/the-secrets-of-the-wave-pilots.html#:~:text=When%20they%20hit%2C%20part%20of,-sight%20%E2%80%94%20these%20and%20other%20patterns.

8 "U.S. Immigration Flows, 1820-2013." Carolina Demography, July 30, 2019. https://carolinademography.cpc.unc.edu/2015/04/27/u-s-immigration-flows-1820-2013/.

9 Bureau, U.S. Census. "Calculating Migration Expectancy Using ACS Data," Census.gov, December 3, 2021. https://www.census.gov/topics/population/migration/guidance/calculating-migration-expectancy.html.

10 Anusha Natarajan, "Key Facts about Recent Trends in Global Migration," Pew Research Center, December 16, 2022. https://www.pewresearch.org/short-reads/2022/12/16/key-facts-about-recent-trends-in-global-migration/.

11 "The Rise of Dual Citizenship: Who Are These Multi-Local Global Citizens?" Global Citizen Forum, January 31, 2022. https://www.globalcitizenforum.org/story/the-rise-of-dual-citizenship-why-multi-local-global-citizens-are-becoming-the-new-normal/#:~:text=Essentially%2C%20anyone%20who%20holds%20two,by%20governments%20around%20the%20world.

12 "UNWTO Tourism Highlights," e-unwto.org, 2018. https://www.e-unwto.org/doi/pdf/10.18111/9789284419876.

13 "Travel & Tourism Economic Impact," World Travel & Tourism Council (WTTC). https://wttc.org/research/economic-impact.

14 Susan C. Antón, Richard Potts, and Leslie C. Aiello, "Evolution of Early Homo: An Integrated Biological Perspective," Science 345 (6192), July 4, 2014. https://doi.org/10.1126/science.1236828.

15 Alan Buis, "Milankovitch (Orbital) Cycles and Their Role in Earth's Climate – Climate Change: Vital Signs of the Planet," NASA, February 7, 2022. https://climate.nasa.gov/news/2948/milankovitch-orbital-cycles-and-their-role-in-earths-climate/.

16 Susan C. Antón, Richard Potts, and Leslie C. Aiello. "Evolution of Early Homo: An Integrated Biological Perspective." See note 14.

17 Jacqueline Armada, "Sustainable Ephemeral: Temporary Spaces with Lasting

Impact," SURFACE at Syracuse University, May 1, 2012. https://surface.syr.edu/honors_capstone/111/.

18 Clay Lancaster, "Metaphysical Beliefs and Architectural Principles," JSTOR, May 1956. https://www.jstor.org/stable/427046.

19 Kevin Nute, *Place, Time and Being in Japanese Architecture*, Psychology Press, 2004. https://philpapers.org/rec/NUTPTA.

20 Tadao Ando, "Laureate Biography," Laureates, The Pritzker Architecture Prize, 1995. https://www.pritzkerprize.com/sites/default/files/file_fields/field_files_inline/1995_bio.pdf.

21 Matsuda Naonori, " Japan's Traditional Houses: The Significance of Spatial Conceptions." Story. In Asia's Old Dwellings: Tradition, Resilience, and Change, 309. Oxford University Press, 2003. https://library.villanova.edu/Find/Record/637894/TOC.

22 Clay Lancaster, "Metaphysical Beliefs and Architectural Principles," JSTOR, May 1956. https://www.jstor.org/stable/427046.

23 "Ecological Threat Register 2021: Understanding Ecological Threats, Resilience, and Peace," Sydney: The Institute for Economics & Peace, October 2021. https://www.economicsandpeace.org/wp-content/uploads/2021/10/ETR-2021-web.pdf.

24 "Life in Za'atari, the Largest Syrian Refugee Camp in the World," *Oxfam International*, May 25, 2022. https://www.oxfam.org/en/life-zaatari-largest-syrian-refugee-camp-world.

25 "Za'atari Refugee Camp – Factsheet, November 2016 – Jordan," ReliefWeb, November 16, 2016. https://reliefweb.int/report/jordan/zaatari-refugee-camp-factsheet-november-2016.

26 Ibid.; Lilly Carlisle, "Jordan's Za'atari Refugee Camp: 10 Facts at 10 Years." UNHCR US, July 2022. https://www.unhcr.org/us/news/stories/jordans-zaatari-refugee-camp-10-facts-10-years; Mario Echeverria, and Moh'd Al-Taher. Jordan: Zaatari Refugee Camp, September 2022. https://www.unhcr.org/jo/wp-content/uploads/sites/60/2022/12/9-Zaatari-Fact-Sheet-September-2022.pdf.

27 "Solving the Housing Challenge of 1.6 Billion People through Sheltertech," Plug and Play Tech Center. https://www.plugandplaytechcenter.com/press/solving-housing-challenges-through-sheltertech/#:~:text=December%2C%2020%2C%202022%20%2D%20If,than%20walls%20and%20a%20roof.

28 Kendall Jeffreys, "Ephemeral Waters." Rachel Carson Council. https://rachelcarsoncouncil.org/ephemeral-waters/.

29 FormsLab, "Additive vs. Subtractive Manufacturing." https://formlabs.com/blog/additive-manufacturing-vs-subtractive-manufacturing/.

30 Rupendra Brahambhatt, "Virginia Is About to 3D-Print an Entire Neighborhood of Homes – and It's Cheaper Than You Think," ZME Science, June 17, 2022. https://www.zmescience.com/ecology/world-problems/3d-printing-houses-17062022/.

31 Tara Massouleh McCay, "Virginia Family Buys First Habitat for Humanity 3D-Printed Home," *Southern Living*, December 29, 2021. https://www.southernliving.com/travel/virginia/virginia-family-buys-first-habitat-for-humanity-3d-printed-home

32 Jessica Cherner, "Habitat for Humanity Debuts First Completed Home Constructed via 3D Printer," *Architectural Digest*, January 3, 2022. https://www.architecturaldigest.com/story/habitat-for-humanity-3d-printer-home.

33 Ibid.
34 "Virginia Launches World's Biggest 3D-Printed Housing Project," *Freethink*, June 11, 2022. https://www.freethink.com/hard-tech/3d-printing-houses#:~:text=Over%20the%20next%205%20years,solve%20America%27s%20affordable%20housing%20crisis.
35 Ibid.
36 "GE Renewable Energy Inaugurates 3D Printing Facility That Will Research More Efficient Ways to Produce Towers for Wind Turbines," *GE News*, April 21, 2022. https://www.ge.com/news/press-releases/ge-renewable-energy-inaugurates-3d-printing-facility-research-more-efficient-ways-produce-towers-for-wind-turbines.
37 James Parkes, "Long-Awaited 3D-Printed Stainless Steel Bridge Opens in Amsterdam," *Dezeen*, July 19, 2021. https://www.dezeen.com/2021/07/19/mx3d-3d-printed-bridge-stainless-steel-amsterdam/.
38 Madeleine Prior, "3D Printed Energy Infrastructure with Lower Material Consumption." 3Dnatives, *3Dnatives*, January 31, 2022. https://www.3dnatives.com/en/3d-printed-energy-infrastructure-with-lower-material-consumption-010220224/.
39 Michael Molitch-Hou, "Has House 3D Printing Finally Made It?" *Forbes Magazine*, June 10, 2022. https://www.forbes.com/sites/michaelmolitch-hou/2022/06/09/has-house-3d-printing-finally-made-it/?sh=12d91748f86a.
40 Ankita Gangotra, Emanuela Del Gado, and Joanna I. Lewis, "3D Printing Has Untapped Potential for Climate Mitigation in the Cement Sector," *Communications Engineering* 2 (6), February 3, 2023. https://doi.org/10.1038/s44172-023-00054-7.
41 Michael Molitch-Hou, "Has House 3D Printing Finally Made It?" See note 39.
42 Paula Pintos, "Tecla Technology and Clay 3D Printed House / Mario Cucinella Architects," *ArchDaily*, April 27, 2021. https://www.archdaily.com/960714/tecla-technology-and-clay-3d-printed-house-mario-cucinella-architects.
43 Adele Peters, "IKEA's 8 principles for circular design show how to build a business based on reuse," *Fast Company*. September 10, 2021. https://www.fastcompany.com/90674372/ikeas-8-principles-for-circular-design-show-how-to-build-a-business-based-on-reuse.

10 Bringing High-Tech Agriculture Indoors

1 Ryan Hobert and Christine Negra, "Climate Change and the Future of Food," United Nations Foundation, September 1, 2020. https://unfoundation.org/blog/post/climate-change-and-the-future-of-food/#:~:text=By%20some%20estimates%2C%20in%20the,the%20brunt%20of%20these%20impacts.
2 "Water Scarcity: Overview," WWF. https://www.worldwildlife.org/threats/water-scarcity.
3 "Hoover Dam," Water Education Foundation. https://www.watereducation.org/aquapedia/hoover-dam; https://www.newsweek.com/lake-mead-water-filling-colorado-explained-reservoir-1783553
4 Dave Davies, "The Colorado River Water Shortage Is Forcing Tough Choices in 7 States," NPR, September 29, 2022. https://www.npr.org/2022/09/29/1125905928/the-colorado-river-water-shortage-is-forcing-

tough-choices-in-7-states; Abrahm Lustgarten and Meridith Kohut, "Climate Change Will Force a New American Migration," *ProPublica*, September 15, 2020. https://www.propublica.org/article/climate-change-will-force-a-new-american-migration.

5 Ken Ritter, "Feds Announce Start of Public Process to Reshape Key Rules on Colorado River Water Use by 2027," *AP News*, June 15, 2023. https://apnews.com/article/colorado-river-water-management-guidelines-drought-d7f09d3e471239d9cafcb4e2dcc53820.

6 Timothy Egan, "The Hoover Dam Made Life in the West Possible. Or So We Thought," *New York Times*, May 14, 2021. https://www.nytimes.com/2021/05/14/opinion/water-hoover-dam-climate-change.html.

7 "Water Facts – Worldwide Water Supply," Bureau of Reclamation, November 4, 2020. https://www.usbr.gov/mp/arwec/water-facts-ww-water-sup.html.

8 David Kirkpatrick, "What Are Vertical Farms, and Can They Really Feed the World?," World Economic Forum, November 30, 2015. https://www.weforum.org/agenda/2015/11/what-are-vertical-farms-and-can-they-really-feed-the-world/.

9 Victoria Masterson, "Vertical Farming – Is This the Future of Agriculture?" Climate Champions, May 24, 2022. https://climatechampions.unfccc.int/vertical-farming-is-this-the-future-of-agriculture/#:~:text=Vertical%20farms%20also%20tend%20to,harvesting%20is%20twice%20a%20year; Team, The Choice. "We Met the Founder of Europe's Largest Vertical Farm," The Choice by ESCP, June 24, 2021. https://thechoice.escp.eu/their-choice/we-met-the-founder-of-europes-largest-vertical-farm/.

10 Antoine Hubert, "Why We Need to Give Insects the Role They Deserve in Our Food Systems," World Economic Forum, July 21, 2021. https://www.weforum.org/agenda/2021/07/why-we-need-to-give-insects-the-role-they-deserve-in-our-food-systems/.

11 Ibid.

12 Arnold van Huis and Dennis G. A. B. Oonincx. "The Environmental Sustainability of Insects as Food and Feed: A Review – Agronomy for Sustainable Development," SpringerLink, September 15, 2017. https://link.springer.com/article/10.1007/s13593-017-0452-8; Jason Plautz, "Eat A Cricket, Save the World," *The Atlantic*, April 27, 2014. https://www.theatlantic.com/politics/archive/2014/04/eat-a-cricket-save-the-world/452844/.

13 Hannah Fuller, "Entomophagy: A New Meaning to 'Tasty Grub,'" Grounded Grub, October 6, 2022. https://groundedgrub.com/articles/entomophagy-a-new-meaning-to-tasty-grub#:~:text=Insects%20are%20also%20cold%2D-blooded,can%20require%20less%20than%202g.

14 "7 Upscale Bug Dishes from around the World," *Food & Wine*, April 21, 2023. https://www.foodandwine.com/travel/gourmet-bug-dishes-around-world.

15 Ibid.

16 Tori Avey, "Discover the History of Sushi," PBS, September 5, 2012. https://www.pbs.org/food/the-history-kitchen/history-of-sushi/#:~:text=Kawafuku%20was%20the%20first%20to,Hollywood%20and%20catered%20to%20celebrities.

11 The Eclipse of Sovereign Nation States and the Gestation of Bioregional Governance

1 Daniel F. Balting, Amir AghaKouchak, Gerrit Lohmann, and Monica Ionita. "Northern Hemisphere Drought Risk in a Warming Climate." Nature News, December 2, 2021. https://www.nature.com/articles/s41612-021-00218-2.
2 Peter Gatrell, "The Nansen Passport: The Innovative Response to the Refugee Crisis That Followed the Russian Revolution," Manchester 1824, February 14, 2019. https://www.manchester.ac.uk/discover/news/the-nansen-passport-the-innovative-response-to-the-refugee-crisis-that-followed-the-russian-revolution/.
3 Gaia Vince, "The Century of Climate Migration: Why We Need to Plan for the Great Upheaval," *Guardian*, August 18, 2022. https://www.theguardian.com/news/2022/aug/18/century-climate-crisis-migration-why-we-need-plan-great-upheaval.
4 Avery Koop, "Ranked: The World's Most and Least Powerful Passports in 2023," *Visual Capitalist*, May 17, 2023. https://www.visualcapitalist.com/most-and-least-powerful-passports-2023/.
5 "The Rise of Dual Citizenship: Who Are These Multi-Local Global Citizens," *Global Citizen Forum*, January 31, 2022. https://www.globalcitizenforum.org/story/the-rise-of-dual-citizenship-why-multi-local-global-citizens-are-becoming-the-new-normal/.
6 Robert Los, "Climate Passport: A Legal Instrument to Protect Climate Migrants – a New Spirit for a Historical Concept", Earth Refuge – The Planet's First Legal Think Tank Dedicated to Climate Migrants, December 31, 2020. https://earthrefuge.org/climate-passport-a-legal-instrument-to-protect-climate-migrants-a-new-spirit-for-a-historical-concept/; Ulrike Grote, Dirk Messner, Sabine Schlacke, and Martina Fromhold-Eisebith, "Just & In-Time Climate Policy: Four Initiatives for a Fair Transformation," German Advisory Council, August 2018.
7 "The 1951 Refugee Convention," UNHCR, 2024. https://www.unhcr.org/about-unhcr/who-we-are/1951-refugee-convention.
8 Robert Los, "Climate Passport". See note 6.
9 Abrahm Lustgarten, "The Great Climate Migration Has Begun," *New York Times*, July 23, 2020. https://www.nytimes.com/interactive/2020/07/23/magazine/climate-migration.html.
10 Xu Chi, Timothy A. Kohler, Timothy M. Lenton, Jens-Christian Svenning, and Marten Scheffer. "Future of the Human Climate Niche," mahb.stanford.edu, October 27, 2019. https://mahb.stanford.edu/wp-content/uploads/2023/12/xu-et-al-2020-future-of-the-human-climate-niche.pdf.
11 Im Eun-Soon, "Deadly Heat Waves Projected in the Densely Populated Agricultural Regions of South Asia," *Science*, August 2, 2017. https://www.science.org/doi/10.1126/sciadv.1603322.
12 Wilfried Ten Brinke, "Permafrost Russia." Climate Change Post. www.climatechangepost.com/russia/permafrost/.
13 Carlos Carroll, "Climatic, Topographic, and Anthropogenic Factors Determine Connectivity," National Library of Medicine, August 1, 2018. https://onlinelibrary.wiley.com/doi/10.1111/gcb.14373.
14 Ibid., 1.
15 Anna Wearn, "Preparing for the Future: How Wildlife Corridors Help Increase

Climate Resilience," Center for Large Landscape Conservation, January 28, 2021. https://largelandscapes.org/news/how-wildlife-corridors-help-increase-climate-resilience/.

16 Brenda Mallory, "Guidance for Federal Departments and Agencies on Ecological Connectivity and Wildlife Corridors," March 21, 2023. https://www.whitehouse.gov/wp-content/uploads/2023/03/230318-Corridors-connectivity-guidance-memo-final-draft-formatted.pdf.

17 Ibid., 1.

18 Ibid., 2.

19 "World's Deadliest Construction Projects: Why Safety Is Important," 360training, January 3, 2023. https://www.360training.com/blog/worlds-deadliest-construction-projects.

20 Charles Maechling, "Pearl Harbor 1941: The First Energy War," *Foreign Service Journal*, August 1979, pp. 11–13.

21 "International Programs – Historical Estimates of World Population," U.S. Census Bureau. https://web.archive.org/web/20130306081718/https://www.census.gov/population/international/data/worldpop/table_history.php.

22 Sébastien Roblin, "The U.S. Military Is Terrified of Climate Change. It's Done More Damage than Iranian Missiles," NBCNews.com. NBCUniversal News Group, September 20, 2020. https://www.nbcnews.com/think/opinion/u-s-military-terrified-climate-change-it-s-done-more-ncna1240484.

23 Andrew Eversden, "'Climate Change Is Going to Cost Us': How the U.S. Military Is Preparing for Harsher Environments," *Defense News*, August 18, 2022. https://www.defensenews.com/smr/energy-and-environment/2021/08/09/climate-change-is-going-to-cost-us-how-the-us-military-is-preparing-for-harsher-environments/.

24 Sébastien Roblin, "The U.S. Military Is Terrified of Climate Change." See note 22.

25 Ibid.

26 Jason Channell et al., "Energy Darwinism II: Why a Low Carbon Future Doesn't Have to Cost the Earth," report, Citi, 2015, p. 8.

27 Patrick Tucker, "Climate Change Is Already Disrupting the Military. It Will Get Worse, Officials Say," Defense One, August 10, 2021. https://www.defenseone.com/technology/2021/08/climate-change-already-disrupting-military-it-will-get-worse-officials-say/184416/.

28 "U.S. Department of Defense – Climate Risk Analysis," Department of Defense. Department of Defense, October 2021. https://media.defense.gov/2021/Oct/21/2002877353/-1/-1/0/DOD-CLIMATE-RISK-ANALYSIS-FINAL.PDF.

29 Department of Defense Climate Adaptation Plan, United States Department of Defense, September 2021, p. 7, https://www.sustainability.gov/pdfs/dod-2021-cap.pdf.

30 "Response to 2017 Hurricanes Harvey, Irma, and Maria: Lessons Learned for Judge Advocates 8," Center for Law & Military Operations, 2018. https://www.loc.gov/rr/frd/Military_Law/pdf/Domestic-Disaster-Response_%20 2017.pdf.

31 Jay Heisler, "World Security Chiefs Debate Military Response to Climate Change," Voice of America (VOA News), November 24, 2021. https://www.voanews.com/a/world-security-chiefs-debate-military-response-to-climate-change-/6326707.html.

12 Two Ways to Listen to the Waters

1 Arthur Schopenhauer, *The World as Will and Representation*, vol. 1, Dover Publications, 1966, pp. 225–226.
2 Ian Sample, "From Bambi to Moby-Dick: How a Small Deer Evolved into the Whale," *Guardian*, December 20, 2007. https://www.theguardian.com/science/2007/dec/20/sciencenews.evolution#:~:text=The%20first%20whales%2C%20Pakicetidae%2C%20emerged,big%20feet%20and%20strong%20tails.
3 Christopher Connery, "There was No More Sea: The Supersession of the Ocean, from the Bible to Cyberspace," *Journal of Historical Geography* 32, 2006: 494–511. 10.1016/j.jhg.2005.10.005.
4 Ibid., 494.
5 Xi Jinping, "Pushing China's Development of an Ecological Civilization to a New Stage," 中国好故事. https://www.chinastory.cn/PCywdbk/english/v1/detail/20190925/1012700000042741569371933649488302_1.html.
6 "China Aims to Build Climate-Resilient Society by 2035," The State Council of the People's Republic of China, June 14, 2022. https://english.www.gov.cn/statecouncil/ministries/202206/14/content_WS62a8342cc6d02e533532c23a.html.
7 Norman MacLean, *Young Men and Fire*, University of Chicago Press, 2017; David Von Drehle, "Opinion | How to Prevent Deadly Wildfires? Stop Fighting Fires," *Washington Post*, September 22, 2022. https://www.washingtonpost.com/opinions/2022/09/22/wildfire-death-prevention-mann-gulch-forest-management/.
8 David von Drehle, "Opinion | How to Prevent Deadly Wildfires? Stop Fighting Fires." *Washington Post*, September 22, 2022. https://www.washingtonpost.com/opinions/2022/09/22/wildfire-death-prevention-mann-gulch-forest-management/.
9 "Study: Third of Big Groundwater Basins in Distress," NASA, Jet Propulsion Laboratory – California Institute of Technology, June 16, 2015. https://www.jpl.nasa.gov/news/study-third-of-big-groundwater-basins-in-distress.
10 Jay Famiglietti, "A Map of the Future of Water," The Pew Charitable Trusts, March 3, 2019. https://www.pewtrusts.org/en/trend/archive/spring-2019/a--map-of-the-future-of-water.
11 "History of Summer Solstice Traditions," National Trust. https://www.nationaltrust.org.uk/discover/history/history-of-summer-solstice-traditions.
12 "Winter Solstice – History," September 21, 2017. https://www.history.com/topics/natural-disasters-and-environment/winter-solstice.
13 Lawrence Wright, *Clockwork Man*, New York: Horizon Press, 1969, p. 47.
14 Eviatar Zerubavel, "Easter and Passover: On Calendars and Group Identity," *American Sociological Review* 47, April 1982: 287–288.
15 Eviatar Zerubavel, "Easter and Passover: On Calendars and Group Identity," *American Sociological Review* 47 (2), April 1982: 288. https://doi.org/10.2307/2094969.
16 Andrew Tarantola, "That Time France Tried to Make Decimal Time a Thing," *Engadget*, January 17, 2022. https://www.engadget.com/that-time-france-tried-to-make-decimal-time-a-thing-143600302.html.

17 Sebastian de Grazia, *Of Time, Work, and Leisure*, New York: Twentieth-Century Fund, 1962, p. 119.
18 Tom Darby, *The Feast: Meditations on Politics and Time*, University of Toronto Press, 1982.
19 Anne Marie Helmenstine, "How Many Molecules Are in a Drop of Water?" *ThoughtCo.*, August 27, 2019. https://www.thoughtco.com/atoms-in-a-drop-of-water-609425.
20 Okoyomon, Adesuwa, "How Long Does the Water Cycle Really Take?" *Science World*, April 15, 2020. https://www.scienceworld.ca/stories/how-long-does-water-cycle-really-take/.

13 Swallowed by the Metaverse or Buoyed by the Aquaverse

1 Josh Howarth, "Alarming Average Screen Time Statistics (2023)," Exploding Topics, January 13, 2023. https://explodingtopics.com/blog/screen-time-stats.
2 Ray Kurzweil, *The Singularity Is Near: When Humans Transcend Biology*, New York: Viking, 2005, p. 30.
3 Ibid., 136.
4 A. M. Turing, "Computing Machinery and Intelligence," *Mind* LIX, issue 236, October 1950: 433–460. https://doi.org/10.1093/mind/LIX.236.433
5 Ethan Siegel, "How Many Atoms Do We Have in Common with One Another?" *Forbes*, April 30, 2020. https://www.forbes.com/sites/startswithabang/2020/04/30/how-many-atoms-do-we-have-in-common-with-one-another/?sh=75adfe6a1b38.
6 J. Gordon Betts et al., *Anatomy and Physiology*, Houston: Rice University, 2013, p. 43; Curt Stager, *Your Atomic Self*, p. 197.
7 David E. Cartwright, *Introduction to Arthur Schopenhauer: On the Basis of Morality*, Providence, RI: Berghahn Books, 1995, p. ix.
8 Ibid.
9 Schopenhauer, *On the Basis of Morality*, p. 61. The quotation within Schopenhauer's quotation is by Immanuel Kant, *Critique of Practical Reason,* Schopenhauer's emphasis.
10 The quotation by Immanuel Kant, *Foundation of the Metaphysics of Morals*, Schopenhauer's emphasis.
11 Schopenhauer, *On the Basis of Morality*, p. 62. See note 9.
12 Ibid., p. 66.
13 Ibid., pp. 143, 147.
14 Ibid., p. 144.
15 Pengfei Li, Jianyi Yang, Mohammad Atiqul Islam, and Shaolei Ren, "Making AI Less "Thirsty": Uncovering and Addressing the Secret Water Footprint of AI Models," *ArXiv* abs/2304.03271, 2023: n.p.
16 Ibid.
17 Sarah Brunswick. "A Tale of Two Shortages: Reconciling Demand for Water and Microchips in Arizona." ABA, February 1, 2023. https://archive.ph/HkL3j#selection-1133.0-1133.15; "Global Semiconductor Sales, Units Shipped Reach All-Time Highs in 2021 as Industry Ramps up Production amid Shortage." Semiconductor Industry Association, February 14, 2022. https://

www.semiconductors.org/global-semiconductor-sales-units-shipped-reach-all-time-highs-in-2021-as-industry-ramps-up-production-amid-shortage/.

18 Pengfei Li, Jianyi Yang, Mohammad Atiqul Islam, and Shaolei Ren, "Making AI Less 'Thirsty.'" See note 15.

19 Bum Jin Park, Yuko Tsunetsugu, Tamami Kasetani, Takahide Kagawa, and Yoshifumi Miyazaki, "The Physiological Effects of *Shinrin-yoku* (Taking in the Forest or Forest Bathing): Evidence from Field Experiments in 24 Forests Across Japan," *Environmental Health and Preventative Medicine* 15 (1), 2010: 21.

20 Roly Russell, Anne D. Guerry, Patricia Balvanera, Rachelle K. Gould, Xavier Basurto, Kai M. A. Chan, Sarah Klain, Jordan Levine, and Jordan Tam. "Humans and Nature: How Knowing and Experiencing Nature Affect Well-Being," *Annual Review of Environment and Resources* 38, October 17, 2013: 473–502. https://doi.org/10.1146/annurev-environ-012312-110838.

21 Andrea Wulf, "A Biography of E. O. Wilson, the Scientist Who Foresaw Our Troubles," *New York Times*, November 10, 2021. https://www.nytimes.com/2021/11/10/books/review/scientist-eo-wilson-richard-rhodes.html; Edward O. Wilson, "The Biological Basis of Morality," *The Atlantic*, April 1998. https://www.theatlantic.com/magazine/archive/1998/04/the-biological-basis-of-morality/377087/.

22 Richard Louv, *Last Child in the Woods*, Chapel Hill, NC: Algonquin Books, 2008.

23 Patrick Barkham, "Should Rivers Have the Same Rights as People?," *Guardian*, July 25, 2021. https://www.theguardian.com/environment/2021/jul/25/rivers-around-the-world-rivers-are-gaining-the-same-legal-rights-as-people.

24 Ibid.

Index